LONDON MATHEMATICAL SOCIETY LECTURE NOTE SERIES

Managing Editor: Professor J.W.S. Cassels, Department of Pure Mathematics and Mathematical Statistics, University of Cambridge, 16 Mill Lane, Cambridge CB2 1SB, England

The books in the series listed below are available from booksellers, or, in case of difficulty, from Cambridge University Press.

4 Algebraic topology, J.F. ADAMS
5 Commutative algebra, J.T. KNIGHT
16 Topics in finite groups, T.M. GAGEN
17 Differential germs and catastrophes, Th. BROCKER & L. LANDER
18 A geometric approach to homology theory, S. BUONCRISTIANO, C.P. ROURKE & B.J. SANDERSON
20 Sheaf theory, B.R. TENNISON
21 Automatic continuity of linear operators, A.M. SINCLAIR
23 Parallelisms of complete designs, P.J. CAMERON
24 The topology of Stiefel manifolds, I.M. JAMES
25 Lie groups and compact groups, J.F. PRICE
27 Skew field constructions, P.M. COHN
29 Pontryagin duality and the structure of LCA groups, S.A. MORRIS
30 Interaction models, N.L. BIGGS
31 Continuous crossed products and type III von Neumann algebras, A. VAN DAELE
34 Representation theory of Lie groups, M.F. ATIYAH *et al*
36 Homological group theory, C.T.C. WALL (ed)
38 Surveys in combinatorics, B. BOLLOBAS (ed)
39 Affine sets and affine groups, D.G. NORTHCOTT
40 Introduction to H_p spaces, P.J. KOOSIS
41 Theory and applications of Hopf bifurcation, B.D. HASSARD, N.D. KAZARINOFF & Y-H. WAN
42 Topics in the theory of group presentations, D.L. JOHNSON
43 Graphs, codes and designs, P.J. CAMERON & J.H. VAN LINT
44 Z/2-homotopy theory, M.C. CRABB
45 Recursion theory: its generalisations and applications, F.R. DRAKE & S.S. WAINER (eds)
46 p-adic analysis: a short course on recent work, N. KOBLITZ
48 Low-dimensional topology, R. BROWN & T.L. THICKSTUN (eds)
49 Finite geometries and designs, P. CAMERON, J.W.P. HIRSCHFELD & D.R. HUGHES (eds)
50 Commutator calculus and groups of homotopy classes, H.J. BAUES
51 Synthetic differential geometry, A. KOCK
52 Combinatorics, H.N.V. TEMPERLEY (ed)
54 Markov processes and related problems of analysis, E.B. DYNKIN
55 Ordered permutation groups, A.M.W. GLASS
56 Journees arithmetiques, J.V. ARMITAGE (ed)
57 Techniques of geometric topology, R.A. FENN
58 Singularities of smooth functions and maps, J.A. MARTINET
59 Applicable differential geometry, M. CRAMPIN & F.A.E. PIRANI
60 Integrable systems, S.P. NOVIKOV *et al*
61 The core model, A. DODD
62 Economics for mathematicians, J.W.S. CASSELS
63 Continuous semigroups in Banach algebras, A.M. SINCLAIR
64 Basic concepts of enriched category theory, G.M. KELLY
65 Several complex variables and complex manifolds I, M.J. FIELD
66 Several complex variables and complex manifolds II, M.J. FIELD
67 Classification problems in ergodic theory, W. PARRY & S. TUNCEL
68 Complex algebraic surfaces, A. BEAUVILLE
69 Representation theory, I.M. GELFAND *et al*
70 Stochastic differential equations on manifolds, K.D. ELWORTHY
71 Groups - St Andrews 1981, C.M. CAMPBELL & E.F. ROBERTSON (eds)
72 Commutative algebra: Durham 1981, R.Y. SHARP (ed)

73 Riemann surfaces: a view towards several complex variables, A.T. HUCKLEBERRY
74 Symmetric designs: an algebraic approach, E.S. LANDER
75 New geometric splittings of classical knots, L. SIEBENMANN & F. BONAHON
76 Spectral theory of linear differential operators and comparison algebras, H.O. CORDES
77 Isolated singular points on complete intersections, E.J.N. LOOIJENGA
78 A primer on Riemann surfaces, A.F. BEARDON
79 Probability, statistics and analysis, J.F.C. KINGMAN & G.E.H. REUTER (eds)
80 Introduction to the representation theory of compact and locally compact groups, A. ROBERT
81 Skew fields, P.K. DRAXL
82 Surveys in combinatorics, E.K. LLOYD (ed)
83 Homogeneous structures on Riemannian manifolds, F. TRICERRI & L. VANHECKE
84 Finite group algebras and their modules, P. LANDROCK
85 Solitons, P.G. DRAZIN
86 Topological topics, I.M. JAMES (ed)
87 Surveys in set theory, A.R.D. MATHIAS (ed)
88 FPF ring theory, C. FAITH & S. PAGE
89 An F-space sampler, N.J. KALTON, N.T. PECK & J.W. ROBERTS
90 Polytopes and symmetry, S.A. ROBERTSON
91 Classgroups of group rings, M.J. TAYLOR
92 Representation of rings over skew fields, A.H. SCHOFIELD
93 Aspects of topology, I.M. JAMES & E.H. KRONHEIMER (eds)
94 Representations of general linear groups, G.D. JAMES
95 Low-dimensional topology 1982, R.A. FENN (ed)
96 Diophantine equations over function fields, R.C. MASON
97 Varieties of constructive mathematics, D.S. BRIDGES & F. RICHMAN
98 Localization in Noetherian rings, A.V. JATEGAONKAR
99 Methods of differential geometry in algebraic topology, M. KAROUBI & C. LERUSTE
100 Stopping time techniques for analysts and probabilists, L. EGGHE
101 Groups and geometry, ROGER C. LYNDON
102 Topology of the automorphism group of a free group, S.M. GERSTEN
103 Surveys in combinatorics 1985, I. ANDERSON (ed)
104 Elliptic structures on 3-manifolds, C.B. THOMAS
105 A local spectral theory for closed operators, I. ERDELYI & WANG SHENGWANG
106 Syzygies, E.G. EVANS & P. GRIFFITH
107 Compactification of Siegel moduli schemes, C-L. CHAI
108 Some topics in graph theory, H.P. YAP
109 Diophantine Analysis, J. LOXTON & A. VAN DER POORTEN (eds)
110 An introduction to surreal numbers, H. GONSHOR
111 Analytical and geometric aspects of hyperbolic space, D.B.A.EPSTEIN (ed)
112 Low-dimensional topology and Kleinian groups, D.B.A. EPSTEIN (ed)
113 Lectures on the asymptotic theory of ideals, D. REES
114 Lectures on Bochner-Riesz means, K.M. DAVIS & Y-C. CHANG
115 An introduction to independence for analysts, H.G. DALES & W.H. WOODIN
116 Representations of algebras, P.J. WEBB (ed)
117 Homotopy theory, E. REES & J.D.S. JONES (eds)
118 Skew linear groups, M. SHIRVANI & B. WEHRFRITZ
119 Triangulated categories in the representation theory of finite-dimensional algebras, D. HAPPEL
120 Lectures on Fermat varieties, T. SHIODA
121 Proceedings of *Groups - St Andrews 1985*, E. ROBERTSON & C. CAMPBELL (eds)
122 Non-classical continuum mechanics, R.J. KNOPS & A.A. LACEY (eds)
123 Surveys in combinatorics 1987, C. WHITEHEAD (ed)
124 Lie groupoids and Lie algebroids in differential geometry, K. MACKENZIE
125 Commutator theory for congruence modular varieties, R. FREESE & R. MCKENZIE
126 Van der Corput's method for exponential sums, S.W. GRAHAM & G. KOLESNIK
127 New directions in dynamical systems, T.J. BEDFORD & J.W. SWIFT (eds)
128 Descriptive set theory and the structure of sets of uniqueness, A.S. KECHRIS & A. LOUVEAU

London Mathematical Society Lecture Note Series. 115

An Introduction to Independence for Analysts

H.G. DALES
School of Mathematics, University of Leeds

W.H. WOODIN
Department of Mathematics, California Institute of Technology

The right of the University of Cambridge to print and sell all manner of books was granted by Henry VIII in 1534. The University has printed and published continuously since 1584.

CAMBRIDGE UNIVERSITY PRESS
Cambridge
New York New Rochelle Melbourne Sydney

CAMBRIDGE UNIVERSITY PRESS
Cambridge, New York, Melbourne, Madrid, Cape Town, Singapore, São Paulo

Cambridge University Press
The Edinburgh Building, Cambridge CB2 8RU, UK

Published in the United States of America by Cambridge University Press, New York

www.cambridge.org
Information on this title: www.cambridge.org/9780521339964

First published 1987
Reprinted 1988
Re-issued in this digitally printed version 2008

A catalogue record for this publication is available from the British Library

ISBN 978-0-521-33996-4 paperback

CONTENTS

PREFACE

The purpose of this book is to explain what it means for a proposition to be independent of set theory, and to describe how independence results can be proved by the technique of forcing. We do this by presenting an application of forcing to a deep and interesting problem in analysis. Our application is, by current standards in set theory, fairly non-technical, and so it offers an excellent setting in which to exhibit to analysts these new techniques from set theory.

Most analysts will have a certain acquaintance with logic and set theory. They will know naïve set theory up to the level of ordinals and cardinals. They will have heard that forcing is a powerful technique that enables one to prove that certain propositions of set theory are independent of specified axioms, and, in particular, that Cohen developed the method of forcing in his proof that the Continuum Hypothesis (CH) is independent of the basic axioms of set theory, ZFC. They may also know of more recently formulated axioms, such as Martin's Axiom (MA), which can be used to establish independence results without the necessity of knowing any of the technicalities of forcing.

However, it is possible that analysts harbour two negative feelings about these matters. First, they may feel that, although logic and set theory are of interest in their own right, they have little to contribute concerning questions which "really" arise in mathematical analysis, and so can be safely left to their disciples. But this is not true. For example, several natural questions about sets of real numbers cannot be resolved in the theory ZFC. For some of these questions the set-theoretic entanglements are quite subtle, in that these questions *can* be resolved by invoking the existence of large cardinal numbers. Here is a specific example. It is not difficult to show that, if $f: \mathbb{R} \to \mathbb{R}$ is a continuous function, then $f(B)$ is Lebesgue measurable for each Borel subset B of $\mathbb{R}$. Now suppose that $f,g : \mathbb{R} \to \mathbb{R}$ are continuous functions. Is $f(\mathbb{R}\setminus g(B))$ Lebesgue measurable for each Borel set B? This problem cannot be decided in ZFC. However if there is a measurable cardinal, then these sets are indeed all Lebesgue measurable.

The main example that we present in this work did arise naturally in analysis. It concerns the automatic continuity of homomorphisms from a Banach algebra of continuous functions into an arbitrary Banach algebra, and the formulation of the problem was such that a solution was expected (perhaps naïvely) in naïve set theory. But eventually it was discovered that the existence of discontinuous homomorphisms from the algebras under consideration could not be decided in ZFC.

Secondly, analysts may feel that the technicalities of forcing are too arcane to be readily accessible. We seek to challenge this view by presenting a reasonably complete account of forcing which is comprehensible to non-logicians: we hope that it will bring them to the point at which they can appreciate the application, which we shall give in detail, of these new ideas to the above automatic continuity problem. We should say, however, that it is not our intention to teach the practical use of forcing in general, but rather to explain quite explicitly how the method of forcing does yield independence and consistency results, and to exemplify this by the study of our chosen example.

Thus this book is directed towards non-logicians, and in particular to analysts. We shall give an account of the background in logic that we shall require, and we shall explain the key notions of proof, of consistency, and of independence. The approach to independence will be through the theory of models: Gödel's completeness theorem allows us to recast independence questions as problems of the construction of models with certain properties.

We shall give full proofs of the results about forcing and Martin's Axiom that we shall need. En route to our main theorem, we shall prove that CH is independent from ZFC, a result that we believe should be known to all mathematicians.

We also hope that our account will be useful for students of set theory as a preliminary to other works. Very little knowledge of analysis is required to follow the details of our example, and the background that is required is given in Chapter 1.

We now describe the problem in analysis that we are to consider. Let X be a compact Hausdorff space, and let $C(X,\mathbb{C})$ be the set of all continuous, complex-valued functions on X. The set $C(X,\mathbb{C})$ is a commutative algebra with respect to the pointwise operations. Set

$$|f|_X = \sup\{|f(x)| : x \in X\} \quad (f \in C(X,\mathbb{C})).$$

Then $|\cdot|_X$ is an algebra norm on $C(X,\mathbb{C})$ - it is the <u>uniform norm</u> - and $(C(X,\mathbb{C}), |\cdot|_X)$ is a Banach algebra. The question we ask is whether or not each algebra norm on $C(X,\mathbb{C})$ is necessarily equivalent to the uniform norm: if so, the topological structure of $C(X,\mathbb{C})$ as a normed algebra would be completely determined by its algebraic structure. This question was first discussed by Kaplansky in 1948, and it was eventually "resolved" independently by Dales and by Esterle in 1976: if the Continuum Hypothesis holds, then, for each infinite compact (Hausdorff) space X, there is an algebra norm on $C(X,\mathbb{C})$ which is not equivalent to the uniform norm.

The appeal to CH in this theorem was thought at the time (at least by those involved) to be an accident and a weakness of the given proofs. However, also in 1976, it was proved by Solovay, using a condition of Woodin, that, if there is a model of set theory, then there is a model of set theory in which each algebra norm on each $C(X,\mathbb{C})$ is equivalent to the uniform norm. Shortly afterwards a different, and easier, approach to the theorem was developed by Woodin; the approach used some techniques of Kunen. It is this approach, which involves Martin's Axiom, that we present here.

These results are included in Woodin's thesis, written at the University of California, Berkeley, and a presentation for experts in forcing is given in the recent volume [45] of Jech, but apart from this ours is the first account of the theorem to appear in print. The original work of Solovay has never been published. A revised version of Woodin's thesis is to appear [72]; this memoir, which will also include several more complicated independence results, is written for logicians, and we can say with quiet confidence that it will be unintelligible to analysts.

Chapter 1 contains only analysis: following the seminal work of Bade and Curtis, we shall analyze the structure of an arbitrary homomorphism from $C(X,\mathbb{C})$ into a Banach algebra. In Chapter 2, we shall discuss partially ordered sets, Boolean algebras, and ultraproducts, topics which form the background to much of our later work, and in Chapter 3 we shall relate our question to one which is amenable to the techniques which are given to us in the theory of forcing: if there is an algebra norm on any $C(X,\mathbb{C})$ which is not equivalent to the uniform norm, then there is a free ultrafilter V on $\mathbb{N}$ and an isotonic map from a subset $\langle Z\rangle/V$ of the ultraproduct $(\mathbb{R}^{\mathbb{N}}/V,<_V)$ into $(\mathbb{N},<_F)$, where $<_F$ is the Fréchet order on $\mathbb{N}$. Also in Chapter 3, we shall discuss the structure of the algebras ℓ^∞/U and c_o/U, where U is a free ultrafilter on $\mathbb{N}$: we formulate some apparently interesting open questions about these algebras.

In Chapter 4, we shall explain how independence can be established by the construction of models. In this chapter, we shall list and discuss the axioms of set theory, and we shall formulate the axiom NDH: "For each compact space X, each homomorphism from $C(X,\mathbb{C})$ into a Banach algebra is continuous.". It is a theorem of Kaplansky, which we discuss in Chapter 1, that for each $C(X,\mathbb{C})$ the uniform norm is the minimal possible algebra norm. Thus the axiom NDH holds if and only if each algebra norm on each $C(X,\mathbb{C})$ is equivalent to the uniform norm.

Chapter 5 first recalls the elementary theory of ordinals and cardinals, and then discusses Martin's Axiom: we give some applications of this axiom to analysis. In Chapter 6, we shall give a final reduction of our problem: we shall prove in ZFC + MA + ¬CH that, if NDH fails, then there is an embedding of a certain totally ordered set $(\mathfrak{R},<)$ into $(\mathbb{N},<_F)$.

In Chapter 7, we shall give our development of forcing: we shall work with Boolean-valued models and with Boolean-valued universes. In our treatment, we place more emphasis than is customary on explaining exactly how forcing arguments yield consistency results. (Let us note for the logician that our treatment is a little non-standard, in that we do not discuss standard models.) At this stage, we prove that CH is independent of ZFC.

Having learnt what forcing is, we must learn how to iterate the process: we study iterated forcing in Chapter 8, essentially proving that "a forcing extension of a forcing extension is a forcing extension". Here again, we are more explicit and give more details than is usual, but we do this at little cost in length. Finally we can conclude the proof of the main theorem of the book: if ZFC is consistent then so is ZFC + MA + NDH.

The practical significance of our main theorem for an analyst is that it is futile to attempt to prove that there is a compact space X and an algebra norm on $C(X,\mathbb{C})$ which is not equivalent to the uniform norm.

We conclude this preface with some remarks about our style of writing.

Throughout, we shall expand points that we guess will be less familiar to analysts (this is a dangerous game for us), and we shall try to use notation which is closer to that traditional in analysis than to that preferred by logicians. But this is not always sensible: logicians do, on occasion, adopt their notation for good reasons, related to the nature of their subject. We have had to find an amicable compromise. For example, let $f : S \to T$ be a function taking the value $f(s)$ at $s \in S$. Now analysts are usually happy to write $f(A)$ for the range of the restriction of f to a subset A of S. But this is dangerously ambiguous, for it may be that S is a transitive set, so that A is also an element of S. In this case, we have followed the analysts, trusting to context to arbitrate the ambiguity.

It is often necessary to be more precise in logic than one would be in analysis. For example, a Boolean algebra is strictly a sextuple

$$\mathcal{B} = \langle B, \wedge, \vee, ', 0, 1 \rangle$$

satisfying certain rules: outside logic, one usually identifies $\mathcal{B}$ with its underlying set B, and we shall do this in Chapters 1 - 6 of this book. However, in Chapters 7 and 8 it seems to be necessary for us to maintain the distinction between $\mathcal{B}$ and B.

Here are some particular notations that we shall use. (An index of notation is given on pages 235-236.) First, $\mathbb{N} = \{1,2,3,\ldots\}$, whereas $\mathbb{Z}^+ = \{0,1,2,\ldots\}$ (so that, again, we are following the analysts). Second, $S \subset T$ allows the possibility that $S = T$. Third, in a phrase such as

$$\text{rad } A = \{a \in A : e - ab \in \text{Inv } A \quad (b \in A)\}$$

the bracket "$(b \in A)$" means "for all b in A".

We give some references in the notes at the end of the chapters to the theorems proved. Although some results are new or have simplified proofs, it is of course not the case that unattributed results are claimed as original.

The authors first discussed this work when the first author visited Caltech in 1984: we are grateful to Professor W. Luxemburg for arranging this visit. We also acknowledge the award of a Visiting Fellowship to the second author by the United Kingdom Science and Engineering Research Council that enabled him to visit Leeds in September, 1984. The second author is an A.P. Sloan Fellow, and we are grateful that this fellowship provided financial support for us to meet at later times, both in the snows of Leeds and in the sunshine of Pacific Palisades, and for us to arrange the typing of this book.

A number of people, both analysts and set theorists, have read some chapters of a preliminary version of this work, and we are very grateful to them for their comments. They include William Bade, John Derrick, Peter Dixon, Peter McClure, Neil Mowbray, John Truss and Tom Ransford. We give special thanks to K.P. Hart and to the referee for reading the entire book in manuscript and for making many valuable suggestions: of course, the decisions on the approach, the (possibly tendentious) opinions, and the errors are all our responsibility.

We owe a special debt of thanks to our typist, Mrs. Joan Bunn. A large amount of information in a work of this type is carried by the details of the symbolism and the fonts. We have tried hard to eliminate errors in this area (presumably unsuccessfully): without the skill and accuracy of Mrs. Bunn's transcription, the process of correcting our errors would surely have diverged.

H.G. Dales and W.H. Woodin,
Leeds, April, 1987

1 HOMOMORPHISMS FROM ALGEBRAS OF CONTINUOUS FUNCTIONS

Throughout this book, we shall work in the naïve set theory familiar to analysts: the formalization of this theory is the axiom system ZFC, which will be discussed in Chapter 4.

We first summarize some elementary facts about Banach algebras that we shall use. The account in the standard text of Rudin [58, Chapters 10 and 11] covers essentially all that we shall require.

All the algebras that we shall consider are linear and associative, and their underlying field is either the complex field $\mathbb{C}$ or the real field $\mathbb{R}$. An algebra is unital if it has an identity; in this case, we often denote the identity by e. The algebra formed by adjoining an identity to a non-unital algebra A is denoted by $A^\#$, and we take $A^\# = A$ if A is unital.

The set of invertible elements of a unital algebra A is denoted by $\mathrm{Inv}\, A$.

A character on a complex algebra A is a non-zero homomorphism from A onto $\mathbb{C}$. The set of characters on A is the character space of A, written Φ_A.

Let A be a commutative algebra. An ideal I of A is modular if the quotient algebra A/I is unital. The (Jacobson) radical, written $\mathrm{rad}\, A$, of A is defined to be the intersection of the maximal modular ideals of A; if A has no such ideals, then $\mathrm{rad}\, A = A$, and in this case A is a radical algebra. (So our convention is that a radical algebra is necessarily commutative.) The algebra A is semisimple if $\mathrm{rad}\, A = \{0\}$.

It is standard that

$$\text{rad } A = \{a \in A : e - ab \in \text{Inv } A^{\#} \quad (b \in A^{\#})\},$$

where e is the identity of $A^{\#}$. Several times we shall use the fact that, if $a \in A$ and $ab = a$ for some $b \in \text{rad } A$, then $a = 0$.

An element a of an algebra A is nilpotent if $a^n = 0$ for some $n \in \mathbb{N}$. We write nil A for the set of nilpotent elements in A. If A is commutative, then nil A is an ideal in A (it is the nilradical of A), and nil $A \subset$ rad A.

An ideal P in a commutative algebra A is a prime ideal if either $a \in P$ or $b \in P$ whenever $a,b \in A$ with $ab \in P$. The algebra A is an integral domain if the zero ideal is a prime ideal in A. We shall occasionally use the standard algebraic fact that, if I is an ideal in A and if $a \in A$ is such that $a^n \notin I$ $(n \in \mathbb{N})$, then there is a prime ideal P with $I \subset P$ and $a \notin P$.

1.1 DEFINITION

Let A be an algebra over a field k. A seminorm on A is a map $p : A \to \mathbb{R}$ such that:

(i) $p(a) \geqslant 0 \quad (a \in A)$;

(ii) $p(\alpha a) = |\alpha| p(a) \quad (\alpha \in k,\ a \in A)$;

(iii) $p(a + b) \leqslant p(a) + p(b) \quad (a,b \in A)$;

(iv) $p(ab) \leqslant p(a)p(b) \quad (a,b \in A)$.

A norm on A is a seminorm p such that

(v) $p(a) \neq 0 \quad (a \in A\backslash\{0\})$.

The algebra A is seminormable (respectively, normable) if there is a non-zero seminorm (respectively, a norm) on A.

It would be more precise to say "algebra seminorm" and "algebra norm", but this extra precision seems to be unnecessary for us.

A norm is usually denoted by $\|\cdot\|$. A Banach

algebra is an algebra with a complete norm. Each normed algebra A is a dense, normed subalgebra of a Banach algebra, the completion of A ([8, 1.12]).

The starting point for the "automatic continuity" theory of Banach algebras is the following basic fact. Let A be a Banach algebra, and let $\phi \in \Phi_A$. Then ϕ is continuous, and in fact

$$\|\phi\| \leqslant 1.$$

The character space Φ_A is a locally compact space with respect to the relative weak *-topology from the dual space of A. (We adopt throughout the convention that a locally compact space is Hausdorff.) If A is a commutative, unital Banach algebra, then Φ_A is compact and non-empty.

Let A be a commutative Banach algebra. If $\Phi_A \neq \emptyset$, then the map $\phi \mapsto \ker \phi$ is a bijection from Φ_A onto the set of maximal modular ideals of A, and so

$$\operatorname{rad} A = \bigcap\{\ker \phi : \phi \in \Phi_A\}. \tag{1}$$

If $\Phi_A = \emptyset$, then A is a radical algebra. Since each character is continuous on A, each maximal modular ideal is closed, and so $\operatorname{rad} A$ is a closed ideal in A.

Let A be an algebra, and let $a \in A$. Then $\sigma(a)$, the spectrum of a, is the set

$$\sigma(a) = \{z \in \mathbb{C} : ze - a \notin \operatorname{Inv} A^{\#}\}.$$

Let A be a Banach algebra. A fundamental theorem ([58, 10.13]) asserts that $\sigma(a)$ is a non-empty, compact subset of $\mathbb{C}$ for each $a \in A$. Set

$$r(a) = \sup\{|z| : z \in \sigma(a)\} \quad (a \in A).$$

Then $r(a)$ is the spectral radius of a, and the spectral radius formula asserts that

$$r(a) = \lim_{n\to\infty} \|a^n\|^{1/n} = \inf \|a^n\|^{1/n} \quad (a \in A).$$

If A is commutative, then $\sigma(a) = \{\phi(a) : \phi \in \Phi_{A^{\#}}\}$, and so $a \in \operatorname{rad} A$ if and only if $r(a) = 0$ ([58, 11.9]). Thus a commutative Banach algebra is a radical algebra if and only if $\|a^n\|^{1/n} \to 0$ as $n \to \infty$ for each $a \in A$.

We now describe in more detail the specific Banach algebras that we are studying.

Let X be a topological space. Then $C(X,\mathbb{C})$ (respectively, $C(X)$) denotes the set of continuous, complex-valued (respectively, real-valued) functions on X. (In making this definition, we implicitly assume that X is non-empty.) Then $C(X,\mathbb{C})$ and $C(X)$ are, respectively, complex and real algebras with respect to the pointwise algebraic operations on X. The constant function 1 is the identity of these algebras.

Let S be a subset of X. For each bounded function f on S, set

$$|f|_S = \sup\{|f(x)| : x \in S\}.$$

Then $|\cdot|_S$ is the <u>uniform norm on</u> S.

Let X be a compact (Hausdorff) space. Then $(C(X,\mathbb{C}), |\cdot|_X)$ is a commutative, unital Banach algebra. For $x \in X$, set $\varepsilon_x(f) = f(x) \quad (f \in C(X,\mathbb{C}))$. Then it is a standard result ([58, 11.13(a)]) that the map $x \mapsto \varepsilon_x$ is a homeomorphism of X onto $\Phi_{C(X,\mathbb{C})}$, and we shall henceforth identify these two spaces.

Let $f \in C(X,\mathbb{C})$, and set $\bar{f}(x) = \overline{f(x)} \quad (x \in X)$, where $\bar{z}$ is the complex conjugate of $z \in \mathbb{C}$. Then $\bar{f} \in C(X,\mathbb{C})$. A subalgebra A of $C(X,\mathbb{C})$ is <u>self-adjoint</u> if $\bar{f} \in A \quad (f \in A)$. We shall use the standard fact ([58, 11.18]) that each closed, self-adjoint subalgebra of $C(X,\mathbb{C})$ which contains 1 is isometrically isomorphic to $C(Y,\mathbb{C})$ for a certain compact space Y.

Let $f \in C(X,\mathbb{C})$. Then $Z(f) = f^{-1}(\{0\})$ is the <u>zero-set of</u> f. A subset F of X is a <u>zero-set</u> if

$F = Z(f)$ for some $f \in C(X,\mathbb{C})$.

Let I be an ideal in $C(X,\mathbb{C})$. The <u>hull</u> of I is the closed subset of X:

$$h(I) = \bigcap\{Z(f) : f \in I\}.$$

Let F be a closed set in X. Then we throughout set

$$I(F) = \{f \in C(X,\mathbb{C}) : Z(f) \supset F\}$$

and

$$J(F) = \{f \in C(X,\mathbb{C}) : Z(f) \text{ is a neighbourhood of } F\}.$$

It is easily seen that $I(F)$ and $J(F)$ are, respectively, the maximal and minimal ideals in $C(X,\mathbb{C})$ whose hull is F, that $I(F)$ is closed, and that $J(F)$ is dense in $I(F)$. Let $x \in X$. Then we write J_x and M_x for $J(\{x\})$ and $I(\{x\})$, respectively. We see that $\{M_x : x \in X\}$ is the family of maximal ideals in $C(X,\mathbb{C})$.

We use the same notations $I(F)$, $J(F)$, J_x and M_x for the analogous ideals in $C(X)$ when we are considering real-valued functions.

We write $\ell^\infty(\mathbb{C})$ (respectively, ℓ^∞) for the set of bounded, complex-valued (respectively, real-valued) sequences on $\mathbb{N}$, and we set

$$|\alpha|_{\mathbb{N}} = \sup\{|\alpha_n| : n \in \mathbb{N}\} \qquad (\alpha = (\alpha_n) \in \ell^\infty(\mathbb{C})).$$

Then $(\ell^\infty(\mathbb{C}), |.|_{\mathbb{N}})$ is a complex, commutative, unital Banach algebra. Temporarily, we denote its character space by Φ. For $n \in \mathbb{N}$, set $\varepsilon_n(\alpha) = \alpha_n$ $(\alpha = (\alpha_n) \in \ell^\infty(\mathbb{C}))$, and, for $\alpha \in \ell^\infty(\mathbb{C})$, set $\hat{\alpha}(\phi) = \phi(\alpha)$ $(\phi \in \Phi)$. The following results are standard and are easily proved; they follow from the Gelfand-Naimark theorem ([58, 11.18]), for example, and are given in greater generality in [35, 8.3].

(i) The map $n \mapsto \varepsilon_n$, $\mathbb{N} \to \Phi$, is a homeomorphism with dense range.

(ii) The map $\alpha \mapsto \hat{\alpha}$, $\ell^\infty(\mathbb{C}) \to C(\Phi,\mathbb{C})$, is an isometric isomorphism.

Thus the compact space Φ has the properties of the <u>Stone-Čech compactification</u> $\beta\mathbb{N}$ of $\mathbb{N}$. In our approach, we define $\beta\mathbb{N}$ to be Φ: different, topological characterizations are given in [36] and [69], for example, and we shall give a characterization in terms of ultrafilters in Chapter 2. Henceforth, we identify $\mathbb{N}$ with a subset of $\beta\mathbb{N}$ and $\ell^\infty(\mathbb{C})$ with $C(\beta\mathbb{N},\mathbb{C})$. For example, if $\mathfrak{p} \in \beta\mathbb{N}$, then $M_\mathfrak{p}$ is a maximal ideal in $\ell^\infty(\mathbb{C})$.

Let σ be a subset of $\mathbb{N}$, and let χ_σ be the characteristic function of σ. Clearly $\{\mathfrak{p} \in \beta\mathbb{N} : \hat{\chi}_\sigma(\mathfrak{p}) = 1\} = \bar{\sigma}$, the closure of σ in $\beta\mathbb{N}$. Thus, if $\{\sigma,\tau\}$ is a partition of $\mathbb{N}$, then $\{\bar{\sigma},\bar{\tau}\}$ is a partition of $\beta\mathbb{N}$ into clopen sets: we shall use this fact several times. It implies that $\overline{\sigma \cap \tau} = \bar{\sigma} \cap \bar{\tau}$ for $\sigma,\tau \subset \mathbb{N}$, and that, for $\mathfrak{p} \in \beta\mathbb{N}$, $\{\bar{\sigma} : \sigma \subset \mathbb{N}, \mathfrak{p} \in \bar{\sigma}\}$ is a base of neighbourhoods of $\mathfrak{p}$ in $\beta\mathbb{N}$ consisting of clopen sets. The existence of this base shows that $\beta\mathbb{N}$ is a totally disconnected space.

We note that, if $\mathfrak{p} \in \beta\mathbb{N}$, then $J_\mathfrak{p}$ is a prime ideal in $\ell^\infty(\mathbb{C})$. For take $f,g \in \ell^\infty(\mathbb{C})$ with $fg \in J_\mathfrak{p}$. Then

$$\mathfrak{p} \in \{n : (fg)(n) = 0\}^-$$

$$= \{n : f(n) = 0\}^- \cup \{n : g(n) = 0\}^-,$$

and so either $\mathfrak{p} \in \{n : f(n) = 0\}^-$, in which case $f \in J_\mathfrak{p}$, or $\mathfrak{p} \in \{n : g(n) = 0\}^-$, in which case $g \in J_\mathfrak{p}$. Thus $J_\mathfrak{p}$ is prime.

We write $c_0(\mathbb{C})$ (respectively, c_0) for the set of complex-valued (respectively, real-valued) sequences which converge to zero. Then $c_0(\mathbb{C})$ is an ideal in $\ell^\infty(\mathbb{C})$, and our identification equates it with the ideal $I(\beta\mathbb{N} \setminus \mathbb{N})$ in $C(\beta\mathbb{N},\mathbb{C})$.

We write $c_{00}(\mathbb{C})$ and c_{00} for the ideals in $c_0(\mathbb{C})$ and c_0, respectively, consisting of sequences (α_n) such that $\alpha_n = 0$ for all but finitely many n.

The question that we shall study in this book is the following.

QUESTION

Let X *be a compact space. Is each norm on the algebra* $C(X,\mathbb{C})$ *necessarily equivalent to the uniform norm* $|\cdot|_X$*?*

The question was raised by Kaplansky in lectures around 1948, and it was Kaplansky who gave the first partial result ([48]).

1.2 THEOREM

Let X be a compact space, and let $\|\cdot\|$ be a norm on $C(X,\mathbb{C})$. Then $|f|_X \leq \|f\| \quad (f \in C(X,\mathbb{C}))$.

<u>Proof</u>

Let A be the completion of the normed algebra $(C(X,\mathbb{C}), \|\cdot\|)$. Then A is a commutative, unital Banach algebra.

The map $\phi \mapsto \phi|C(X,\mathbb{C})$, $\Phi_A \to \Phi_{C(X,\mathbb{C})}$, is a continuous injection, and so we may regard Φ_A as a closed subspace of X. Suppose, if possible, that $\Phi_A \neq X$, and take $x \in X\backslash\Phi_A$. Then there exist f and g in $C(X,\mathbb{C})$ such that $f(\Phi_A) = g(\Phi_A) = \{0\}$, such that $g(x) = 1$, and such that $fg = g$. By (1), $f \in \text{rad}\, A$, and so $g = 0$, a contradiction. Thus $\Phi_A = X$.

Let $f \in C(X,\mathbb{C})$, and let $x \in X$. Since $x \in \Phi_A$, $|f(x)| \leq \|f\|$, and so $|f|_X \leq \|f\|$, as required. ∎

It follows immediately from 1.2 (via the open mapping theorem) that each <u>complete</u> norm on $C(X,\mathbb{C})$ is equivalent to the uniform norm, and so there is an inequivalent norm on $C(X,\mathbb{C})$ if and only if there is an incomplete norm.

Let $\|\cdot\|$ be an incomplete norm on $C(X,\mathbb{C})$, and let A be the completion of $(C(X,\mathbb{C}), \|\cdot\|)$. Then the embedding $C(X,\mathbb{C}) \to A$ is a discontinuous monomorphism. On the other hand, suppose that θ is a discontinuous homomorphism from $C(X,\mathbb{C})$ into a Banach algebra. Set

$$\|f\| = \max\{|f|_X, \|\theta(f)\|\} \quad (f \in C(X,\mathbb{C})).$$

Then $\|.\|$ is an incomplete norm on $C(X,\mathbb{C})$. Thus our question is equivalent to the question of the existence of a discontinuous homomorphism from $C(X,\mathbb{C})$ into a Banach algebra, and it will usually be convenient to formulate it in this way.

The next advance in the study of norms on $C(X,\mathbb{C})$ after the work of Kaplansky was the seminal paper of Bade and Curtis of 1960 ([2]). In this paper, the structure of an arbitrary homomorphism from $C(X,\mathbb{C})$ into a Banach algebra was analysed, and the key notions of singularity set and radical homomorphism (see below) were introduced. Later, in 1976, Johnson ([47]) extended Bade and Curtis's theorem, and used the extension to show that, if there is an incomplete norm on $C(X,\mathbb{C})$ for any compact space X, then there exists $\mathfrak{p} \in \beta\mathbb{N}\setminus\mathbb{N}$, a radical Banach algebra R, and a non-zero homomorphism θ from $c_o(\mathbb{C})$ into R such that $\ker\theta \supset J_{\mathfrak{p}}$. It is this reduction theorem that we shall need in Chapter 3. Our proof is slightly more direct, and gives a little more information, than that obtained by following the original route.

The main technical device in the proof of Bade and Curtis's theorem is the following <u>main boundedness theorem</u>, which originates in [2, Theorem 2.1].

1.3 THEOREM

Let A and B be Banach algebras, and let θ be a homomorphism from A into B. Suppose that there are sequences (a_n) and (b_n) in A such that $a_m b_n = 0$ $(m,n \in \mathbb{N},\ m \neq n)$. Then there exists a constant C such that

$$\|\theta(a_n b_n)\| \leq C\|a_n\|\,\|b_n\| \quad (n \in \mathbb{N}).$$

Proof

We can suppose that $\|a_n\| = \|b_n\| = 1 \quad (n \in \mathbb{N})$.

Suppose, if possible, that the result is false. Then for each $(i,j) \in \mathbb{N} \times \mathbb{N}$, we may choose $n(i,j) \in \mathbb{N}$ such that the map $(i,j) \mapsto n(i,j)$ is injective and such that

$$\|\theta(u_{ij}v_{ij})\| \geq 4^{i+j} \quad (i,j \in \mathbb{N}),$$

where $u_{ij} = a_{n(i,j)}$ and $v_{ij} = b_{n(i,j)}$. Set

$$f_i = \sum_{\ell=1}^{\infty} v_{i\ell}/2^{\ell} \quad (i \in \mathbb{N}),$$

so that each $f_i \in A$. Choose $j(i) \in \mathbb{N}$ so that

$$\|\theta(f_i)\| \leq 2^{j(i)} \quad (i \in \mathbb{N}),$$

and set

$$g = \sum_{k=1}^{\infty} u_{k,j(k)}/2^k,$$

so that $g \in A$. Now, for $i \in \mathbb{N}$,

$$gf_i = \sum_{k=1}^{\infty}\sum_{\ell=1}^{\infty} u_{k,j(k)}v_{i\ell}/2^{k+\ell}$$

$$= u_{i,j(i)}v_{i,j(i)}/2^{i+j(i)}$$

and so

$$\|\theta(gf_i)\| \geq 4^{i+j(i)}/2^{i+j(i)} = 2^{i+j(i)}.$$

But

$$\|\theta(gf_i)\| \leq \|\theta(g)\|\,\|\theta(f_i)\| \leq 2^{j(i)}\,\|\theta(g)\|,$$

and so $\|\theta(g)\| \geq 2^i$ for all $i \in \mathbb{N}$, a contradiction.

Thus the result holds. ∎

We introduce some further notation. Let X be a compact space. We write N_x for the family of open neighbourhoods of a point x of X. If U is an open subset of X, then $K_U = I(X\setminus U)$, so that K_U consists of those functions of $C(X,\mathbb{C})$ which vanish off U, and K_U is a closed ideal in $C(X,\mathbb{C})$. For example, $K_\emptyset = \{0\}$.

1.4 DEFINITION

Let θ be a homomorphism from $C(X,\mathbb{C})$ into a Banach algebra. A point $x \in X$ is a <u>singularity point</u> for θ if, for each $U \in N_x$, $\theta|K_U$ is discontinuous. The set of singularity points is the <u>singularity set</u> for θ.

1.5 DEFINITION

A <u>radical homomorphism at x</u> is a non-zero homomorphism from a maximal ideal M_x of $C(X,\mathbb{C})$ into the radical of a commutative Banach algebra.

Let $x \in X$, and let $\theta : M_x \to A$ be a homomorphism into a commutative Banach algebra A. We *claim* that θ is a radical homomorphism if and only if $\theta|J_x = 0$. For suppose that $\theta(M_x) \subset \operatorname{rad} A$, and take $f \in J_x$. Then there exists $g \in M_x$ with $fg = f$, and $\theta(f)\theta(g) = \theta(f)$, so that $\theta(f) = 0$. Thus $\theta|J_x = 0$. Conversely, suppose that $\theta|J_x = 0$, and take $\phi \in \Phi_A$. Then $\phi\circ\theta \in \Phi_{M_x} \cup \{0\}$, and $(\phi\circ\theta)|J_x = 0$. Since J_x is dense in M_x, $\phi\circ\theta = 0$, and so $\theta(M_x) \subset \ker\phi$. Thus $\theta(M_x) \subset \operatorname{rad} A$, and θ is a radical homomorphism. The claim is established.

It is further clear that there is a radical homomorphism at x if and only if M_x/J_x is seminormable.

We shall several times use the elementary fact that, if F is an infinite subset of a regular space X, then there exist $x_n \in F$ and $U_n \in N_{x_n}$ for $n \in \mathbb{N}$ such that $U_m \cap U_n = \emptyset$ $(m \neq n)$. For take $x,y \in F$ with $x \neq y$. Then there exist $U \in N_x$ and $V \in N_y$ with $x \notin \bar{V}$, $y \notin \bar{U}$, and $U \cup V = X$. Either $U \cap F$ or $V \cap F$ is infinite, and so we can choose $x_1 \in F$ and an open set W_1 such that

$W_1 \cap F$ is infinite and $x_1 \notin \bar{W}_1$. For $n = 2,3,\ldots$, successively choose $x_n \in W_1 \cap \ldots \cap W_{n-1} \cap F$ and an open set W_n such that $W_1 \cap \ldots \cap W_n \cap F$ is infinite and $x_n \notin \bar{W}_n$. Set $U_1 = X \setminus \bar{W}_1$ and set

$$U_n = (W_1 \cap \ldots \cap W_{n-1}) \setminus \bar{W}_n \quad (n = 2,3,\ldots).$$

There is one final remark before we prove the theorem of Bade and Curtis. Let A be a Banach algebra, let $\theta : C(X,\mathbb{C}) \to A$ be a homomorphism and let I be an ideal of $C(X,\mathbb{C})$ of the form K_U or $K_U \cap J_x$. Suppose that there is a constant k such that $\|\theta(g^2)\| \leq k|g|_X^2 \quad (g \in I)$. Take $f \in I$. Then $f = f_1 - f_2 + i(f_3 - f_4)$, where $f_j \in I$ and $f_j(X) \subset [0, |f|_X] \quad (j = 1,\ldots,4)$. Further, there exist $g_j \in I$ with $g_j^2 = f_j$, and so

$$\|\theta(f_j)\| \leq k|f_j|_X \quad (j = 1,\ldots,4).$$

Thus $\|\theta(f)\| \leq 4k|f|_X$. Hence, if $\theta|I$ is discontinuous, then, for each $n \in \mathbb{N}$, there exists $g_n \in I$ with $\|\theta(g_n^2)\| > n|g_n|_X^2$. Similarly, we see that, if $\theta|I \neq 0$, then there exists $g \in I$ such that $\theta(g^2) \neq 0$.

We can now give the theorem of Bade and Curtis ([2, Theorem 4.3]).

1.6 THEOREM

Let X be a compact space, let θ be a homomorphism from $C(X,\mathbb{C})$ into a Banach algebra A, and let F be the singularity set of θ. Then:

(i) F is empty if and only if θ is continuous;

(ii) F is finite.

Now suppose that θ is discontinuous and that $F = \{x_1,\ldots,x_n\}$.

(iii) There are a continuous homomorphism $\mu : C(X,\mathbb{C}) \to A$ and linear maps $\nu_1,\ldots,\nu_n : C(X,\mathbb{C}) \to A$ such that $\nu_i|M_{x_i}$ is a radical homomorphism at x_i $(i = 1,\ldots,n)$ and $\theta = \mu + \nu_1 + \ldots + \nu_n$.

Proof

(i) Clearly $F = \emptyset$ if θ is continuous. Conversely, suppose that $F = \emptyset$. Then, for each $x \in X$, there exists $U_x \in N_x$ such that $\theta|K_{U_x}$ is continuous, and so X has a finite open cover V such that $\theta|K_V$ is continuous for each $V \in V$. Using a partition of unity subordinate to V, we see easily that θ is continuous.

(ii) Suppose, if possible, that F is infinite. Then there are an infinite sequence $(x_n) \subset F$ and $U_n \in N_{x_n}$ such that $U_m \cap U_n = \emptyset$ $(m \neq n)$. Choose $f_n \in K_{U_n}$ such that $\|\theta(f_n^2)\| > n|f_n|_X^2$ $(n \in \mathbb{N})$. Since $f_m f_n = 0$ $(m \neq n)$, this contradicts the main boundedness theorem. Thus F is finite.

(iii) Let $x \in F$. We *claim* that there exists $U_x \in N_x$ such that $\theta|(K_{U_x} \cap J_x)$ is continuous. For suppose that this is not the case. Then we may inductively choose sequences $(V_n) \subset N_x$ and $(f_n) \subset J_x$ such that, for each $n \in \mathbb{N}$, $f_n \in K_{V_n}$, $f_i(V_{n+1}) = \{0\}$ $(i = 1,\ldots,n)$, and $\|\theta(f_n^2)\| > n|f_n|_X^2$. Since $f_m f_n = 0$ $(m \neq n)$, this again contradicts the main boundedness theorem. Thus the claim holds.

It is immediate from the claim that $\theta|J(F)$ is continuous, say $\|\theta(f)\| \leqslant k|f|_X$ $(f \in J(F))$.

Fix $W_i \in N_{x_i}$ $(i = 1,\ldots,n)$ such that $W_i \cap W_j = \emptyset$ $(i \neq j)$, and fix $e_i \in K_{W_i}$ such that $|e_i|_X = 1$ and $e_i = 1$ on some neighbourhood of x_i. Let B be the set of functions $f \in C(X,\mathbb{C})$ such that $f - f(x_i)1 \in J_{x_i}$ $(i = 1,\ldots,n)$. Then B is a dense subalgebra of $C(X,\mathbb{C})$. For $f \in B$, $f - \sum_{i=1}^{n} f(x_i)e_i$ belongs to $J(F)$, and so

$$\|\theta(f)\| \leqslant k\left|f - \sum_{i=1}^{n} f(x_i)e_i\right|_X + \left\|\sum_{i=1}^{n} f(x_i)\theta(e_i)\right\|$$

$$\leqslant \left[(n+1)k + \sum_{i=1}^{n}\|\theta(e_i)\|\right]|f|_X.$$

Thus θ is continuous on B.

Let μ be the continuous extension of $\theta|B$ to $C(X,\mathbb{C})$, and let $\nu = \theta - \mu$. Then $\mu : C(X,\mathbb{C}) \to A$ is a continuous homomorphism, and $\nu|J(F) = 0$.

Take $f \in I(F)$ and $g \in J(F)$. Then

$$\theta(f)\mu(g) = \theta(fg) = \mu(fg) = \mu(f)\mu(g),$$

and so

$$\nu(f)\mu(g) = (\theta(f) - \mu(f))\mu(g) = 0.$$

Since μ is continuous and $\overline{J(F)} = I(F)$, we have $\nu(f)\mu(g) = 0 \quad (f,g \in I(F))$. Thus, for $f,g \in I(F)$,

$$\begin{aligned}\nu(fg) &= \theta(f)\theta(g) - \mu(f)\mu(g)\\ &= (\mu(f) + \nu(f))(\mu(g) + \nu(g)) - \mu(f)\mu(g)\\ &= \nu(f)\nu(g),\end{aligned}$$

and so $\nu|I(F)$ is a homomorphism.

Set $\nu_i(f) = \nu(fe_i) \quad (f \in C(X,\mathbb{C}),\ i = 1,\ldots,n)$. Then $\nu_1,\ldots,\nu_n : C(X,\mathbb{C}) \to A$ are linear maps. Since $1 - \sum_{i=1}^{n} e_i \in J(F)$, $\nu = \nu_1 + \ldots + \nu_n$. Let $i \in \{1,\ldots,n\}$ and let $f,g \in M_{x_i}$. Then $fe_i, ge_i \in I(F)$, and so

$$\nu(fe_i)\nu(ge_i) = \nu(fge_i^2).$$

But $e_i^2 - e_i \in J(F)$, and so $\nu(fge_i^2) = \nu(fge_i)$. Thus $\nu_i(f)\nu_i(g) = \nu_i(fg)$, and $\nu_i : M_{x_i} \to A$ is a homomorphism. If $f \in J_{x_i}$, then $fe_i \in J(F)$, and so $\nu_i(f) = 0$. Thus $\nu_i|M_{x_i}$ is a radical homomorphism. ∎

1.7 COROLLARY

Let X be a compact space. Then the following are equivalent:

(a) there is a norm on $C(X,\mathbb{C})$ which is not

equivalent to the uniform norm;

(b) there is a discontinuous homomorphism from $C(X,\mathbb{C})$ into a Banach algebra;

(c) there is a radical homomorphism from a maximal ideal of $C(X,\mathbb{C})$;

(d) there exists $x \in X$ such that M_x/J_x is seminormable. ∎

1.8 THEOREM

Let X be a compact space. Assume that there is a discontinuous homomorphism from $C(X,\mathbb{C})$ into a Banach algebra. Then there exists $p \in \beta\mathbb{N}\setminus\mathbb{N}$, a radical Banach algebra R, and a non-zero homomorphism θ from $c_o(\mathbb{C})$ into R such that $\ker\theta \supset J_p \cap c_o(\mathbb{C})$.

Proof

Let θ be a discontinuous homomorphism from $C(X,\mathbb{C})$ into a Banach algebra A. Then there exists $(f_n) \subset C(X,\mathbb{C})$ such that $|f_n|_X = 1$ and $\|\theta(f_n)\| \to \infty$. Let B be the smallest closed, unital subalgebra of $C(X,\mathbb{C})$ containing $\{f_n, \bar{f}_n : n \in \mathbb{N}\}$. Then B is a closed, separable, self-adjoint subalgebra of $C(X,\mathbb{C})$, and so B is isometrically isomorphic to $C(Y,\mathbb{C})$ for a certain compact space Y. Since B is separable, Y is metrizable, and clearly $\theta|B$ is discontinuous.

By 1.6, there is a singularity point $x \in Y$, a radical Banach algebra R, and a non-zero homomorphism $\nu : M_x \to R$ with $\nu|J_x = 0$. Certainly x is not an isolated point of Y. Set $X_o = Y\setminus\{x\}$.

Let (U_n) be a sequence of open subsets of X_o such that $U_m \cap U_n = \emptyset$ $(m \neq n)$, and let

$$S = \{n \in \mathbb{N} : \nu|K_{U_n} \neq 0\}.$$

For each $n \in S$, there exists $g_n \in K_{U_n}$ such that $\nu(g_n^2) \neq 0$. Since $K_{U_n} \cap J_x$ is dense in K_{U_n}, and since $\nu|J_x = 0$, we may suppose that $\|\nu(g_n^2)\| > n|g_n|_Y^2$. Since

$g_m g_n = 0 \quad (m \neq n)$, it follows from the main boundedness theorem that S is finite.

Let d be a metric defining the topology on Y, and take $(x_n) \subset X_o$ such that $\delta_n \to 0$ and $4\delta_{n+1} < \delta_n \quad (n \in \mathbb{N})$, where $\delta_n = d(x,x_n)$. Set

$$V_n = \{y \in X_o : \tfrac{1}{2}\delta_n < d(y,x) < 2\delta_n\} \quad (n \in \mathbb{N}).$$

The sets V_n are non-empty and open, and $V_m \cap V_n = \emptyset \quad (m \neq n)$. Let $\{\sigma_m : m \in \mathbb{N}\}$ be a countable partition of $\mathbb{N}$ into infinite subsets, and let $U_m = \bigcup\{V_n : n \in \sigma_m\}$. Then (U_m) is a sequence of open subsets of X_o such that $U_m \cap U_n = \emptyset \quad (m \neq n)$, and so $\nu|K_{U_m} = 0$ for all but finitely many m. By passing to a subsequence of (x_n), we may suppose that $\nu|K_V = 0$, where $V = \bigcup\{V_n : n \in \mathbb{N}\}$.

For $n \in \mathbb{N}$, set

$$F_n = \{y \in X_o : 2\delta_{n+1} \leqslant d(y,x) \leqslant \tfrac{1}{2}\delta_n\}.$$

Suppose that $F_n = \emptyset$ for all but finitely many n. Then $V \cup \{x\}$ is a neighbourhood of x in Y, and so, if $f \in M_x$, there exists $f_1 \in J_x$ and $f_2 \in K_V$ with $f = f_1 + f_2$, whence $\nu(f) = \nu(f_1) + \nu(f_2) = 0$, a contradiction. Thus we may suppose that $F_n \neq \emptyset \quad (n \in \mathbb{N})$. Let $F = \bigcup F_n \cup \{x\}$, a closed set in Y.

For $n \in \mathbb{N}$, take $e_n \in J_x$ with $|e_n|_Y = 1$, with $e_n(F_n) = \{1\}$, and with $e_n(y) = 0$ if $d(y,x) > \delta_n$ or $d(y,x) < \delta_{n+1}$. Set

$$\psi(\alpha) = \sum_{n=1}^{\infty} \alpha_n e_n \quad (\alpha = (\alpha_n) \in c_o(\mathbb{C})).$$

For each $y \in X_o$, there exists $W \in N_y$ with $e_n(W) = 0$ for all but two values of n, and so $\psi(\alpha)$ is continuous on X_o. Clearly, $\psi(\alpha) \in M_x$, and $\nu \circ \psi : c_o(\mathbb{C}) \to R$ is a linear map. For each $y \in X_o \backslash V$, $e_n(y) \in \{0,1\}$, and so $e_n^2 - e_n \in K_V$ and $\nu(e_n^2 - e_n) = 0 \quad (n \in \mathbb{N})$. Also

$e_m e_n = 0 \quad (m \neq n)$, and so $\nu\circ\psi$ is a homomorphism.

Since $\psi(c_{oo}(\mathbb{C})) \subset J_x \subset \ker \nu$, $(\nu\circ\psi)\, c_{oo}(\mathbb{C}) = 0$.

Choose $g \in M_x$ with $\nu(g) \neq 0$, and let $\beta_n = |g|_{F_n} \quad (n \in \mathbb{N})$, so that $(\beta_n) \in c_o$. Take $\alpha_n > 0$ such that $\alpha = (\alpha_n) \in c_o$ and $(\beta_n/\alpha_n) \in c_o$, set $f = g/\alpha_n$ on F_n, and set $f(x) = 0$. Then f is continuous on the closed set F in Y, and so f extends to a function in M_x. Since $\psi(\alpha)f = g$ on F, $(\nu\circ\psi)(\alpha)\nu(f) = \nu(g)$, and so $(\nu\circ\psi)(\alpha) \neq 0$. Thus $\nu\circ\psi : c_o(\mathbb{C}) \to R$ is a non-zero homomorphism with $(\nu\circ\psi)|c_{oo}(\mathbb{C}) = 0$.

We now regard $c_o(\mathbb{C})$ as an ideal in $C(\beta\mathbb{N}, \mathbb{C})$. Set

$$E = \{\mathfrak{p} \in \beta\mathbb{N} : (\nu\circ\psi)|(K_U \cap c_o(\mathbb{C})) \neq 0 \quad (U \in N_{\mathfrak{p}})\}.$$

Since $\nu\circ\psi \neq 0$, $E \neq \emptyset$. Since $(\nu\circ\psi)|c_{oo}(\mathbb{C}) = 0$, $E \subset \beta\mathbb{N}\setminus\mathbb{N}$. If E were infinite, there would exist a sequence (U_n) of non-empty, open subsets of $\beta\mathbb{N}$ such that $U_m \cap U_n = \emptyset \quad (m \neq n)$ and $(\nu\circ\psi)|(K_{U_n} \cap c_o(\mathbb{C})) \neq 0 \quad (n \in \mathbb{N})$. Since $(\nu\circ\psi)|c_{oo}(\mathbb{C}) = 0$, the argument of the third paragraph leads to a contradiction. Thus E is finite. Clearly, $(\nu\circ\psi)|(J(E) \cap c_o(\mathbb{C})) = 0$. Let $\mathfrak{p} \in E$, and take $e \in C(\beta\mathbb{N}, \mathbb{C})$ such that $e = 1$ on a neighbourhood of $\mathfrak{p}$ and $e = 0$ on a neighbourhood of $E\setminus\{\mathfrak{p}\}$. Set $\theta(f) = (\nu\circ\psi)(fe) \quad (f \in c_o(\mathbb{C}))$. Then θ is the required homomorphism. ∎

Bade and Curtis left open the question of the existence of discontinuous homomorphisms from the algebras $C(X,\mathbb{C})$ into Banach algebras. Eventually, such homomorphisms were constructed for each infinite, compact space by Dales ([16]) and by Esterle ([26], [24], [27]) in 1976. However, both constructions required the assumption of the Continuum Hypothesis (CH), and are thus theorems of the system ZFC + CH. Neither Dales nor Esterle discussed whether or not CH was required.

We now state without proof the results of Dales and of Esterle. We denote the cardinality of a set S by $|S|$;

a review of cardinal numbers will be given in Chapter 5.

Theorems whose proof is given in the theory ZFC + CH (see Chapter 4) are labelled "CH", but theorems which hold in ZFC are not labelled.

1.9 THEOREM (CH)

Let X be a compact space, and let P be a non-maximal, prime ideal in $C(X,\mathbb{C})$ such that $|C(X,\mathbb{C})/P| = 2^{\aleph_0}$. Then the algebra $C(X,\mathbb{C})/P$ is normable. ∎

Now, if X is infinite, then $C(X,\mathbb{C})$ does contain non-maximal, prime ideals P such that $|C(X,\mathbb{C})/P| = 2^{\aleph_0}$. Further, it is easily seen that, if P is any prime ideal in $C(X,\mathbb{C})$, then the hull of P is a singleton, and so there exists $x \in X$ such that $J_x \subset P \subset M_x$; since $\bar{J}_x = M_x$, P is closed if and only if P is equal to the maximal ideal M_x. Thus it follows from 1.9 that, if the Continuum Hypothesis holds, then the four properties (a) - (d) of Corollary 1.7 are true for each infinite, compact space X. In particular, we have the following result.

1.10 THEOREM (CH)

Let X be an infinite compact space. Then there exists a discontinuous homomorphism from $C(X,\mathbb{C})$ into a Banach algebra, and there is a norm on $C(X,\mathbb{C})$ which is not equivalent to the uniform norm.

Let X be a compact space. If X is separable, then $|C(X,\mathbb{C})| = 2^{\aleph_0}$. Suppose that $x \in X$ is such that $M_x \neq J_x$ (so that x is not a P-point of X - see Definition 2.22), and take $f \in M_x \backslash J_x$. Then $f^n \notin J_x$ $(n \in \mathbb{N})$, and so there is a prime ideal P in $C(X,\mathbb{C})$ with $J_x \subset P$ and $f \notin P$. Hence we have two further results.

1.11 THEOREM (CH)

Let X be a separable, infinite compact space.

(i) Each non-maximal, prime ideal of $C(X,\mathbb{C})$ is the kernel of a discontinuous homomorphism into a Banach algebra.

(ii) If $x \in X$ and if $M_x \neq J_x$, then the quotient algebra M_x/J_x is seminormable.

1.12 **THEOREM** (CH)

Let $\mathfrak{p} \in \beta\mathbb{N}\backslash\mathbb{N}$. Then the algebras $\ell^\infty(\mathbb{C})/J_\mathfrak{p}$ and $c_o(\mathbb{C})/(J_\mathfrak{p} \cap c_o(\mathbb{C}))$ are normable.

We have stated the above four theorems as results of ZFC + CH. In fact, they all follow from a more general result on the normability of integral domains that can be proved in ZFC. This more general result will be discussed at the end of Chapter 6.

It is an immediate consequence of Theorem 1.8 that, if there is a discontinuous homomorphism from $C(X,\mathbb{C})$ for any compact space X, then there is a discontinuous homomorphism from $c_o(\mathbb{C})$. It is natural to enquire whether or not all infinite compact spaces are equivalent for our problem. The first result in this direction is the following theorem.

1.13 **THEOREM**

Assume that there is a discontinuous homomorphism from $\ell^\infty(\mathbb{C})$ into a Banach algebra. Then there is a discontinuous homomorphism from $C(X,\mathbb{C})$ into a Banach algebra for each infinite compact space X.

<u>Proof</u>

Since X is infinite, it contains an infinite, discrete subspace $\{x_n\}$. For $f \in C(X,\mathbb{C})$, set

$$\tau(f)(n) = f(x_n) \quad (n \in \mathbb{N}).$$

Then $\tau(f) \in \ell^\infty(\mathbb{C})$, and $\tau : C(X,\mathbb{C}) \to \ell^\infty(\mathbb{C})$ is a continuous homomorphism with range B, say. For each $\alpha \in B$ and each

$\varepsilon > 0$, there exists $f \in C(X,\mathbb{C})$ with $\tau(f) = \alpha$ and $|f|_X \leqslant |\alpha|_{\mathbb{N}} + \varepsilon$, and so B is a closed subalgebra of $\ell^\infty(\mathbb{C})$.

Since there is a discontinuous homomorphism from $\ell^\infty(\mathbb{C})$, there exists $p \in \beta\mathbb{N}\setminus\mathbb{N}$ and a non-zero homomorphism θ from $\ell^\infty(\mathbb{C})$ into a Banach algebra A such that $\theta|M_p \neq 0$ and $\theta|J_p = 0$.

Take $f \in M_p$ with $\theta(f^2) \neq 0$, choose $(k_n) \subset \mathbb{N}$ such that $0 < |f(k_{n+1})| < |f(k_n)| < 1/n$ $(n \in \mathbb{N})$, and, for $n \in \mathbb{N}$, take

$$U_n = \{m \in \mathbb{N} : |f(k_{n+1})| < |f(m)| \leqslant |f(k_n)|\}.$$

Let $U_0 = \mathbb{N}\setminus \bigcup U_n$. Then $\{U_0, U_1, \ldots\}$ is a partition of $\mathbb{N}$. Set $\gamma_m = |f(k_n)|$ for $m \in U_n$ $(n \in \mathbb{Z}^+)$, taking $\gamma_m = 1$ if $m \in U_0$. Then $(\gamma_m) \in \ell^\infty$, and so (γ_m) has a continuous extension, say g, to $\beta\mathbb{N}$. Clearly, $|f(x)| \leqslant g(x)$ $(x \in \beta\mathbb{N})$ and $g \in M_p$. Set $h(x) = f^2(x)/g(x)$ if $g(x) \neq 0$, and $h(x) = 0$ if $g(x) = 0$. Then $|h(x)| \leqslant |f(x)| \leqslant g(x)$ $(x \in \beta\mathbb{N})$, and so $h \in M_p$. Since $gh = f^2$, $\theta(g) \neq 0$.

For $\alpha = (\alpha_n) \in B$, set

$$\psi(\alpha) = \sum_{n=0}^{\infty} \alpha_n \chi_n,$$

where χ_n is the characteristic function of U_n. Then $\psi : B \to \ell^\infty(\mathbb{C})$ and $\theta\circ\psi\circ\tau : C(X,\mathbb{C}) \to A$ are homomorphisms.

If $\alpha \in c_{00}(\mathbb{C})$, then $\psi(\alpha) \in J_p$, and so $(\theta\circ\psi)|c_{00}(\mathbb{C}) = 0$. Let $\beta = (|f(k_n)|)$. Then $\beta \in c_0(\mathbb{C})$ and $\psi(\beta) = \gamma$. Since $\theta(g) \neq 0$, $(\theta\circ\psi)(\beta) \neq 0$, and so $\theta\circ\psi\circ\tau$ is discontinuous. ∎

It follows from 1.8 and 1.13 that all infinite compact spaces are equivalent for our problem if and only if the existence of a discontinuous homomorphism from $c_0(\mathbb{C})$ implies the existence of one from $\ell^\infty(\mathbb{C})$. We do not know whether or not this is true, and the relation between $\ell^\infty(\mathbb{C})$ and $c_0(\mathbb{C})$ in this regard seems to be subtle. We shall return to this point again in Chapter 3.

1.14 NOTES

Fuller accounts of Banach algebra theory are given in [8] and [17]. Automatic continuity theory - the study of conditions under which linear maps from Banach algebras are automatically continuous - is described in [65], [51], and [15], as well as [17]. Generalizations of the main boundedness theorem are given in these sources.

As we stated, the notions of singularity set and radical homomorphism were introduced by Bade and Curtis in [2]. The theory of singularity points has been developed in much more general contexts: see, for example, [1] and [51].

It is a striking fact that, in distinction to the commutative case, all homomorphisms from certain non-commutative C^*-algebras (such as $\mathcal{B}(H)$, the algebra of all bounded linear operators on a Hilbert space H) into a Banach algebra *are* automatically continuous: this is a theorem of Johnson ([46], [65, 12.4], [17]).

The concentration in Theorem 1.9 on prime ideals is natural, for the following result is proved in [65, 11.4]:

> Let θ be a discontinuous homomorphism from $C(X,\mathbb{C})$ onto a dense subalgebra of a Banach algebra A. Then there is a closed ideal J in A such that the map $f \mapsto \theta(F) + J$, $C(X,\mathbb{C}) \to A/J$, is a discontinuous homomorphism whose kernel is a prime ideal in $C(X,\mathbb{C})$.

Let X be an infinite compact space such that $|C(X,\mathbb{C})| = 2^{\aleph_0}$, and let I be an ideal in $C(X,\mathbb{C})$. Then, with CH, $C(X,\mathbb{C})/I$ is normable whenever I is a finite intersection of prime ideals. It is not known whether or not the quotient $C(X,\mathbb{C})/I$ is ever normable for any other ideal: if $\ell^\infty(\mathbb{C})/I$ is normable, then I is a finite intersection of prime ideals ([25, Proposition 5.1]). Other partial results are given in [18].

It is not true, even with CH, that $C(X,\mathbb{C})/P$ is normable for each prime ideal P: this may fail if $|C(X,\mathbb{C})/P| > 2^{\aleph_0}$ ([25, Théorème 7.1], [17]). Indeed it may be that the normability of $C(X,\mathbb{C})/P$ entails that $|C(X,\mathbb{C})/P| = 2^{\aleph_0}$: this is an open question that we shall return to in Chapter 6.

The characterization of radical Banach algebras R such that, with CH, there is a discontinuous homomorphism from $C(X,\mathbb{C})$ into $R^{\#}$ has been established by Esterle ([30], [17, §6.6]): R has this property if and only if there exists $a \in R\backslash\{0\}$ such that $a \in \overline{a^2R}$. Several other characterizations of R are also given in [17, §6.6]. Here are two examples of such radical Banach algebras.

Let H_0^∞ denote the set of bounded analytic functions f on the open right-hand half-plane Π such that $f(z) \to 0$ as $z \to \infty$ in Π. Then H_0^∞ is a Banach algebra with respect to the uniform norm, and $e^{-z}H_0^\infty$ is a closed ideal in H_0^∞. The quotient algebra $H_0^\infty/e^{-z}H_0^\infty$ is a radical Banach algebra with the required property ([17, §6.5]).

Let R be the set of Lebesgue integrable functions f on $\mathbb{R}^+$ such that

$$\|f\| = \int_0^\infty |f(t)|e^{-t^2}\,dt < \infty.$$

For $f,g \in R$, set

$$(f * g)(t) = \int_0^t f(t-s)g(s)ds \quad (t \in \mathbb{R}^+).$$

Then $(R,*,\|\cdot\|)$ is a radical Banach algebra with the required property ([17, §6.5]). It follows from Titchmarsh's convolution theorem that R is an integral domain.

It is a theorem of Esterle ([29], [17]) that each epimorphism from each $C(X,\mathbb{C})$ onto a Banach algebra is automatically continuous.

2 PARTIAL ORDERS, BOOLEAN ALGEBRAS, AND ULTRAPRODUCTS

Our approach to forcing will use "Boolean-valued universes", a method which originates with Scott and Solovay (see the notes to Chapter 7). It is therefore necessary for us to present some basic facts about Boolean algebras; we shall prove only the results that we shall require later. In fact, our discussion will be carried out in the more general context of partially ordered sets. In Example 2.2(vi) we shall describe $(\mathbb{N}^{\mathbb{N}},<_F)$ and $(\mathbb{N}^{\mathbb{N}},<<_F)$, examples of partially ordered sets which will be very prominent in later chapters.

In this chapter, we shall introduce the <u>completion</u> and the <u>Stone space</u> of a Boolean algebra B. The latter is the set of ultrafilters on B, and we shall recognize $\beta\mathbb{N}$ as the Stone space of the Boolean algebra $\mathcal{P}(\mathbb{N})$. This allows us to reformulate our basic question as a question about $c_o/\mathcal{U}$, where $\mathcal{U}$ is a free ultrafilter on $\mathbb{N}$.

We shall conclude the chapter by discussing the P-points of the compact space $\beta\mathbb{N}\setminus\mathbb{N}$.

2.1 DEFINITION

Let P be a non-empty set. A <u>strict partial order</u> on P is a binary relation $<$ such that:

(i) if $a < b$ and $b < c$ in P, then $a < c$;

(ii) $a \not< a$ for each $a \in P$.

A <u>total order</u> on P is a strict partial order $<$ such that:

(iii) for each $a,b \in P$, either $a < b$ or $a = b$ or $b < a$.

A partially ordered set is a pair $\mathcal{P} = (P,<)$, where P is a non-empty set and $<$ is a strict partial order on P.

A subset Q of a partially ordered set $(P,<)$ is a chain if the restriction of $<$ to $Q \times Q$ is a total order on Q.

The first coordinate P of the pair $\mathcal{P}$ is the underlying set of the partially ordered set. For the present we shall not distinguish notationally between a partially ordered set and its underlying set.

Let $(P,<)$ be a partially ordered set. Set $a \leqslant b$ in P if $a < b$ or $a = b$. The relation $\leqslant$ satisfies the conditions: (i) if $a \leqslant b$ and $b \leqslant c$ in P, then $a \leqslant c$; (ii) $a \leqslant a$ for each $a \in P$; (iii) if $a \leqslant b$ and $b \leqslant a$ in P, then $a = b$. A relation satisfying these conditions is a partial order. Now let $\leqslant$ be a partial order on a non-empty set P, and set $a < b$ if $a \leqslant b$ and $a \neq b$ $(a,b \in P)$. Then $(P,<)$ is a partially ordered set. *We shall always suppose that the orders* $<$ *and* $\leqslant$ *correspond in this way*, without further mention.

Let S be a non-empty set. We write $\mathcal{P}(S)$ for the power set of S: $\mathcal{P}(S)$ is the set of all subsets of S.

2.2 EXAMPLES

(i) Let S be a non-empty set. For $a,b \in \mathcal{P}(S)$, set $a \leqslant b$ if $a \subset b$. Then $\leqslant$ is a partial order on $\mathcal{P}(S)$. This is the ordering given by inclusion.

(ii) For $a,b \in \mathcal{P}(\mathbb{N})$, set $a <_F b$ if $a \backslash b$ is finite and if $b \backslash a$ is infinite. Then $<_F$ is a strict partial order on $\mathcal{P}(\mathbb{N})$. The order $<_F$ is inclusion modulo the finite sets.

(iii) Let A be a radical algebra. For $a,b \in A \backslash \{0\}$, set $a \ll b$ if there exists $c \in A$ with $a = bc$. Then $\ll$ is a strict partial order on $A \backslash \{0\}$: condition 2.1(ii) is satisfied, for if $a = ac$ in A, then $a = 0$. The order $\ll$ is the divisibility order on A.

(iv) Let X be a topological space. For $f, g \in C(X)$, set $f \leqslant g$ if $f(x) \leqslant g(x)$ $(x \in X)$. Then $\leqslant$ is a partial order on $C(X)$: it is the <u>standard order</u>.

(v) Let $(P_1,<),\dots,(P_n,<)$ be totally ordered sets, and let $P = P_1 \times \dots \times P_n$. Set $(x_1,\dots,x_n) < (y_1,\dots,y_n)$ in P if $x_i < y_i$, where $i = \min\{j : x_j \neq y_j\}$. Then $<$ is a total order on P, called the <u>lexicographic order</u>.

(vi) Let $f, g \in \mathbb{N}^{\mathbb{N}}$. Then $f(n) < g(n)$ <u>eventually</u> if there exists $n_o \in \mathbb{N}$ such that $f(n) < g(n)$ $(n \geqslant n_o)$. Set

$$f <_F g \quad \text{if} \quad f(n) < g(n) \quad \text{eventually},$$

$$f <<_F g \quad \text{if} \quad g(n) - f(n) \to \infty \quad \text{as} \quad n \to \infty.$$

Then $<_F$ and $<<_F$ are strict partial orders on $\mathbb{N}^{\mathbb{N}}$: they are the <u>Fréchet order</u> and the <u>strong Fréchet order</u>, respectively.

2.3 DEFINITION

Let $(P,<)$ and $(Q,<)$ be partially ordered sets. A map π from P to Q is:

(i) <u>order-preserving</u> if $\pi(a) \leqslant \pi(b)$ whenever $a \leqslant b$ in P;

(ii) <u>isotonic</u> if $\pi(a) < \pi(b)$ whenever $a < b$ in P;

(iii) <u>anti-isotonic</u> if $\pi(a) > \pi(b)$ whenever $a < b$ in P;

(iv) an <u>embedding</u> if π is injective and if $\pi(a) \leqslant \pi(b)$ in Q if and only if $a \leqslant b$ in P;

(v) an <u>order-isomorphism</u> if π is a surjective embedding.

Note that, if π is injective, then π is order-preserving if and only if it is isotonic. Also, if $<$ is a total order on P and if $\pi : P \to Q$ is isotonic, then π is an embedding.

The following result concerning the Fréchet order on $\mathbb{N}^{\mathbb{N}}$ will be required later.

2.4 PROPOSITION

There is an anti-isotonic map from $(\mathbb{N}^{\mathbb{N}}, <_F)$ into $(\mathbb{N}^{\mathbb{N}}, <_F)$.

<u>Proof</u>

Let $\gamma : \mathbb{N} \times \mathbb{N} \to \mathbb{N}$ be a fixed bijection.
For $f \in \mathbb{N}^{\mathbb{N}}$, set

$$S(f) = \{(m,n) \in \mathbb{N} \times \mathbb{N} : f(m) < n\},$$

so that $S(f)$ is the set of points "above" the graph of f, and set

$$\pi(f)(n) = |\{1,\ldots,n\} \cap \gamma(S(f))| + 1 \quad (n \in \mathbb{N}).$$

Then $\pi(f) \in \mathbb{N}^{\mathbb{N}}$.

We *claim* that $\pi : (\mathbb{N}^{\mathbb{N}}, <_F) \to (\mathbb{N}^{\mathbb{N}}, <_F)$ is an anti-isotonic map. For suppose that $f <_F g$ in $\mathbb{N}^{\mathbb{N}}$. Then there exists $m_o \in \mathbb{N}$ such that $f(m) < g(m) \quad (m \geq m_o)$. Since

$$S(g)\backslash S(f) \subset \{(m,n) : n \leq f(m),\ 1 \leq m < m_o\},$$

$S(g)\backslash S(f)$ is finite, and, since $(m,f(m)) \in S(f)\backslash S(g)$ $(m \geq m_o)$, $S(f)\backslash S(g)$ is infinite. Thus $\pi(f)(n) > \pi(g)(n)$ eventually, and so $\pi(g) <_F \pi(f)$ in $\mathbb{N}^{\mathbb{N}}$. ∎

Let $(P,<)$ be a partially ordered set, let Q be a non-empty subset of P, and let $a \in P$. Then:

(i) a is a <u>maximal</u> element of Q if $a \in Q$ and if $a \not< b \quad (b \in Q)$;

(ii) a is a <u>maximum</u> element of Q if $a \in Q$ and if $b \leq a \quad (b \in Q)$;

(iii) a is an upper bound of Q if $b \leqslant a$ $(b \in Q)$;

(iv) a is a supremum of Q if a is an upper bound of Q and if $a \leqslant b$ for each upper bound b of Q.

Minimal, minimum, lower bound, and infimum are defined similarly.

The supremum and infimum of Q are necessarily unique if they exist; they are denoted by sup Q and inf Q, respectively. Note that, if $(P,<)$ is totally ordered, then a maximal element of Q is the maximum of Q.

Zorn's Lemma is the assertion: if each chain in a partially set P has an upper bound, then P has a maximal element. The relation of Zorn's Lemma to the Axiom of Choice will be recalled in Chapter 5. However, as is usual in analysis, we shall use the lemma freely.

2.5 DEFINITION

A partially ordered set is complete if each non-empty subset which has an upper bound has a supremum and if each non-empty subset which has a lower bound has an infimum.

2.6 DEFINITION

A Boolean algebra is a structure

$$\mathcal{B} = (B,\wedge,\vee,',0,1),$$

where $\wedge$ and $\vee$ are binary operations on the set B, where $'$ is a unary operation on B, where 0 and 1 are distinct elements of B, and where the following hold for all $a,b,c \in B$:

(i) $a \wedge b = b \wedge a$, $a \vee b = b \vee a$;

(ii) $a \wedge 1 = a \vee 0 = a$;

(iii) $a \wedge a = a \vee a = a$;

(iv) $a \wedge (b \vee c) = (a \wedge b) \vee (a \wedge c)$;

(v) $a \wedge (b \wedge c) = (a \wedge b) \wedge c$;

(vi) $(a \wedge b)' = a' \vee b'$;

(vii) $(a')' = a$;

(viii) $a \vee a' = 1,\ a \wedge a' = 0$;

(ix) $0' = 1$.

The above list contains redundancies: see [39, §2]. The operations $\wedge$, $\vee$, and $'$ are called meet, join and complementation, respectively.

The first coordinate of the sextuple $(B,\wedge,\vee,',0,1)$ is the underlying set of the Boolean algebra $\mathcal{B}$. At present we shall use the letter B for both the Boolean algebra and for its underlying set. A Boolean algebra has cardinality κ if its underlying set has cardinality κ.

A non-empty subset of a Boolean algebra is a subalgebra if it is a Boolean algebra with respect to the given operations. The set $\{0,1\}$ is a Boolean algebra, to be denoted by (2): it is the trivial Boolean algebra, and it is regarded as a subalgebra of each Boolean algebra.

Let B and C be Boolean algebras. A map π from B to C is a Boolean homomorphism if, for $a,b \in B$,

$$\pi(a \wedge b) = \pi(a) \wedge \pi(b), \quad \pi(a \vee b) = \pi(a) \vee \pi(b),$$
$$\pi(a') = \pi(a)'.$$

An injective (respectively, bijective) Boolean homomorphism is a Boolean monomorphism (respectively, Boolean isomorphism). Write $B \cong C$ if there is an isomorphism from B onto C.

Let B be a Boolean algebra. For $a,b \in B$, set $a \leqslant_B b$ if $a \wedge b = a$. Then $(B,<_B)$ is easily checked to be a partially ordered set, where $<_B$ is the strict partial order corresponding to $\leqslant_B$. In fact, it is not difficult to see that each of the operations $\wedge$, $\vee$, and $'$ can be defined in terms of the relation $\leqslant_B$, and so a Boolean algebra is just a partially ordered set in which the order satisfies certain additional axioms.

A Boolean algebra B is complete if the partially

ordered set $(B, <_B)$ is complete, and so, in a complete Boolean algebra, each non-empty subset S has a supremum and an infimum: they are denoted by

$$\bigvee\{s : s \in S\} \quad \text{or} \quad \bigvee S,$$

and

$$\bigwedge\{s : s \in S\} \quad \text{or} \quad \bigwedge S,$$

respectively. For $a \in B$, we have

$$a \wedge \bigvee\{s : s \in S\} = \bigvee\{a \wedge s : s \in S\},$$

for example.

Here is an obvious example of a complete Boolean algebra. Let S be a non-empty set, and let $B = P(S)$. For $a, b \in B$, set

$$a \wedge b = a \cap b, \quad a \vee b = a \cup b, \quad a' = S \backslash a, \qquad (1)$$

and set $0 = \emptyset$, $1 = S$. Then B is a Boolean algebra, and the corresponding order is ordering by inclusion. The supremum and infimum of a non-empty subset A of B are $\bigcup\{a : a \in A\}$ and $\bigcap\{a : a \in A\}$, respectively. When we refer to $P(S)$ as a Boolean algebra, we suppose that it has the Boolean operations given by (1). We shall show, in Corollary 2.15, that the algebras $P(S)$ are universal in the class of Boolean algebras. First, however, we give another important example of a Boolean algebra.

2.7 <u>EXAMPLE</u>

Let X be a non-empty topological space which is not necessarily Hausdorff. A subset U of X is <u>regular-open</u> if it is the interior of its closure: $U = \text{int}(\bar{U})$. For example, for each open set U, $\text{int}(\bar{U})$ is regular-open. Let $R(X)$ be the set of all regular-open subsets of X. For $U, V \in R(X)$, set

$$U \wedge V = U \cap V, \quad U \vee V = \mathrm{int}(\overline{U \cup V}),$$
$$U' = \mathrm{int}(X \setminus U), \tag{2}$$

and, as before, set $0 = \emptyset$, $1 = X$. It is routine, but not entirely trivial, to check that, with respect to these operations, $R(X)$ is a complete Boolean algebra (cf. [39, §4], [69, §2.3]): we have

$$\bigvee\{U : U \in S\} = \mathrm{int}\overline{\bigcup\{U : U \in S\}},$$
$$\bigwedge\{U : U \in S\} = \mathrm{int}\bigcap\{U : U \in S\}, \tag{3}$$

for a non-empty subset S of $R(X)$. The Boolean algebra $R(X)$ is the <u>regular-open algebra</u> of X. Note that, if $U,V \in R(X)$, then $U \leqslant_{R(X)} V$ if and only if $U \subset V$.

Let $B(X)$ be the family of clopen subsets of X. Then $B(X)$ is a subset of $R(X)$, and the restrictions of the Boolean operations given by (2) to $B(X)$ agree with the Boolean operations given by (1). Thus $B(X)$ is a Boolean subalgebra of both $R(X)$ and $P(X)$.

The space X is <u>extremely disconnected</u> if the closure of each open subset of X is open. In this case, we see that $R(X) = B(X)$, and so $R(X)$ is a complete Boolean algebra with respect to the operations given by (1).

2.8 DEFINITION

Let $(P,<)$ be a partially ordered set.

(i) Two elements a and b of P are <u>compatible</u> if there exists $c \in P$ with $c \leqslant a$ and $c \leqslant b$, and a and b are <u>incompatible</u> if they are not compatible.

(ii) A subset Q of P is <u>dense</u> in P if, for each $a \in P$, there exists $b \in Q$ with $b \leqslant a$.

We write $a \perp b$ if a and b are incompatible elements of P.

These notions will be applied to a Boolean algebra B by regarding $(B\backslash\{0\},<_B)$ as a partially ordered set. Thus, if $a,b \in B\backslash\{0\}$, then $a \perp b$ if and only if $a \wedge b = 0$; for example, elements a and b in $\mathcal{P}(S)$ are incompatible if and only if they are non-empty and disjoint. We stress that we consider $B\backslash\{0\}$, and not B itself.

We shall now associate to each partially ordered set a complete Boolean algebra.

2.9 DEFINITION

Let $(P,<)$ be a partially ordered set, and let B be a complete Boolean algebra. Then (B,π) is a <u>completion</u> of P if $\pi : (P,<) \to (B\backslash\{0\},<_B)$ is a map such that:

(i) $\pi(P)$ is dense in $B\backslash\{0\}$;

(ii) π is order-preserving;

(iii) for each $a,b \in P$, $a \perp b$ in P if and only if $\pi(a) \perp \pi(b)$ in $B\backslash\{0\}$.

2.10 THEOREM

Let $(P,<)$ be a partially ordered set. Then P has a completion.

<u>Proof</u>

For $a \in P$, let $U_a = \{b \in P : b \leqslant a\}$. We see that $a \in U_a$ $(a \in P)$, and that, if $c \in U_a \cap U_b$, then $c \in U_c \subset U_a \cap U_b$. Thus $\{U_a : a \in P\}$ is a base for a topology, say τ, on P.

Let $B = \mathcal{R}(P)$, the regular-open algebra of (P,τ). Then B is a complete Boolean algebra. Set

$$\pi : a \mapsto \operatorname{int}(\bar{U}_a), \quad P \to B\backslash\{0\}.$$

We check that conditions (i) - (iii) of Definition 2.9 are satisfied.

(i) Take $a \in B\backslash\{0\}$. Then a is a non-empty, open set in (P,τ), and so there exists $b \in P$ with

$U_b \subset a$. We have $\pi(b) \subset \text{int}\,\bar{a} = a$, and so $\pi(b) \leqslant_B a$.

(ii) If $a \leqslant b$ in P, then $U_a \subset U_b$, and so $\pi(a) \leqslant_B \pi(b)$.

(iii) Let a and b be compatible in P, and take $c \in P$ with $c \leqslant a$ and $c \leqslant b$. Then $\pi(c) \in B\backslash\{0\}$, $\pi(c) \leqslant_B \pi(a)$ and $\pi(c) \leqslant_B \pi(b)$, and so $\pi(a)$ and $\pi(b)$ are compatible in $B\backslash\{0\}$.

Now let a and b be incompatible in P, so that $U_a \cap U_b = \emptyset$. Since U_b is open, $\bar{U}_a \cap U_b = \emptyset$, and so $\pi(a) \cap U_b = \emptyset$. Since $\pi(a)$ is open, $\pi(a) \cap \bar{U}_b = \emptyset$, and so $\pi(a) \cap \pi(b) = \emptyset$. Thus $\pi(a)$ and $\pi(b)$ are incompatible in $B\backslash\{0\}$. ∎

It can be shown that a completion of a partially ordered set P is unique, in the sense that, if (B_1,π_1) and (B_2,π_2) are completions of P, then there is a Boolean isomorphism $\theta : B_1 \to B_2$ such that $\pi_2 = \theta\circ\pi_1$. If (B,π) is a completion of P, we set $B = \bar{P}$.

It is not claimed in Theorem 2.10 that $\pi : P \to B$ is always an injection. For example, suppose that P has a minimum element a. Then $a \in U_b$ for each $b \in P$, and so $\bar{U}_b = P$ $(b \in P)$. Thus $R(P) = \{\emptyset,P\}$, and B is the trivial Boolean algebra.

However, it is easily seen that $\pi : P \to B$ is an injection if P is <u>separative</u>, in the sense that, for each $a,b \in P$ with $b \nleqslant a$ there exists $c \in P$ with $c \perp a$ and $c \leqslant b$.

2.11 DEFINITION

Let $(P,<)$ be a partially ordered set. A subset F of P is a <u>prefilter</u> if, for each $a,b \in F$, there exists $c \in F$ with $c \leqslant a$ and $c \leqslant b$.

A prefilter F is a <u>filter</u> if, for each $a \in F$ and $b \in P$ with $a \leqslant b$, it follows that $b \in F$.

A filter is a <u>maximal filter</u> if it is maximal in the set of filters on P with respect to inclusion.

Let F be a prefilter in P. Then

$$\{b \in P : a \leqslant b \text{ for some } a \in F\}$$

is a filter in P containing F. It is immediate from Zorn's Lemma that each filter is contained in a maximal filter, and so each prefilter is contained in a maximal filter.

Let B be a Boolean algebra. Then a <u>proper filter</u> in B is a filter in the partially ordered set $(B\backslash\{0\},<_B)$, and an <u>ultrafilter</u> in B is a maximal filter in this set. Thus a proper filter in B is a subset A such that: (i) if $a,b \in A$, then $a \wedge b \in A$; (ii) if $a \in A$ and $a \leqslant_B b$, then $b \in A$; (iii) $0 \notin A$. It is easily checked that a proper filter A is an ultrafilter if and only if, for each $a \in B\backslash\{0\}$, either a or a' belongs to A. Let A be a subset of B, and consider the set

$$C = \{a_1 \wedge \ldots \wedge a_n : a_1,\ldots,a_n \in A\}.$$

If $0 \notin C$, then C is a prefilter in $B\backslash\{0\}$, and so, in this case, A is contained in an ultrafilter in B.

Let S be a non-empty set. Then proper filters and ultrafilters in the Boolean algebra $P(S)$ are said to be <u>filters</u> and <u>ultrafilters on S</u>, respectively. Thus, a filter on S is a family F of subsets such that: (i) if $F,G \in F$, then $F \cap G \in F$; (ii) if $F \in F$ and $F \subset G$, then $G \in F$; (iii) $\emptyset \notin F$. For example, if x belongs to a topological space X, then the family of neighbourhoods of x in X is a filter on X. Note that, if U is an ultrafilter on S, and if $A,B \subset S$ with $A \cup B \in U$, then either $A \in U$ or $B \in U$.

Let B be a Boolean algebra. We know that the partially ordered set $(B\backslash\{0\},<_B)$ has a completion. We now give a different, perhaps more familiar, construction of the completion in this special case.

The family of ultrafilters in B is denoted by $\mathcal{S}(B)$. Set

$$\pi(b) = \{\mathcal{U} \in \mathcal{S}(B) : b \in \mathcal{U}\} \quad (b \in B). \tag{4}$$

Certainly, $\pi(0) = \emptyset$ and $\pi(b) \neq \emptyset$ $(b \in B\backslash\{0\})$. Also, for $a, b \in B$, $\pi(a \wedge b) = \pi(a) \cap \pi(b)$, $\pi(a \vee b) = \pi(a) \cup \pi(b)$, and $\pi(a') = \mathcal{S}(B)\backslash\pi(a)$. In particular,

$$\{\pi(b) : b \in B\}$$

is a base for a topology, say τ_B, on $\mathcal{S}(B)$.

2.12 DEFINITION

The topology τ_B is the <u>Stone topology</u>, and $(\mathcal{S}(B), \tau_B)$ is the <u>Stone space</u> of B.

2.13 LEMMA

(i) Each set $\pi(b)$ is clopen in $(\mathcal{S}(B), \tau_B)$.

(ii) $(\mathcal{S}(B), \tau_B)$ is a compact, totally disconnected space.

(iii) If D is a dense subset of $B\backslash\{0\}$, then $\bigcup\{\pi(d) : d \in D\}$ is a dense, open set in $\mathcal{S}(B)$.

<u>Proof</u>

(i) By definition, each $\pi(b)$ is open. Since $\mathcal{S}(B)\backslash\pi(b) = \pi(b')$, each $\pi(b)$ is closed.

(ii) If $\mathcal{U} \neq \mathcal{V}$ in $\mathcal{S}(B)$, there exists $b \in \mathcal{U}\backslash\mathcal{V}$, and $\pi(b)$ is a clopen set such that $\mathcal{U} \in \pi(b)$ and $\mathcal{V} \notin \pi(b)$. Thus $\mathcal{S}(B)$ is totally disconnected (and Hausdorff).

To show that $(\mathcal{S}(B), \tau_B)$ is compact, it suffices to show that, if $\{\pi(a) : a \in A\}$ has the finite intersection property, where A is a subset of B, then $\bigcap\{\pi(a) : a \in A\}$ is non-empty. But in this case $a_1 \wedge \ldots \wedge a_n \neq 0$ $(a_1, \ldots, a_n \in A)$, and so A is contained in an ultra-filter, say $\mathcal{U}$, on B. Since $\mathcal{U} \in \bigcap\{\pi(a) : a \in A\}$, the result follows.

(iii) Let $U = \bigcup\{\pi(d) : d \in D\}$. Certainly U is open in $\mathcal{S}(B)$. For each $b \in B\setminus\{0\}$, there exists $d \in D$ with $d \leqslant b$. Since $\pi(d) \leqslant \pi(b)$, $\pi(b) \cap U \neq \emptyset$, and so U is dense in $\mathcal{S}(B)$. ∎

2.14 THEOREM

Let B be a Boolean algebra, and let $\bar{B} = \mathcal{R}(\mathcal{S}(B))$. Then $(\bar{B},\pi)$ is a completion of $(B\setminus\{0\},<_B)$, and $\pi : B \to \bar{B}$ is a Boolean monomorphism.

Proof

By 2.13(i), $\pi(B) \subset \mathcal{B}(\mathcal{S}(B)) \subset \bar{B}$, and π is a Boolean homomorphism. Take $a,b \in B$ with $a \neq b$, say $a \not\leqslant_B b$. Then $a \wedge b' \neq 0$, and so there is an ultrafilter $\mathcal{U}$ in B containing a and b'; $a \in \mathcal{U}$ and $b \notin \mathcal{U}$, and so $\pi(a) \neq \pi(b)$. Thus π is an injection.

Take $a \in \bar{B}\setminus\{0\}$. Then a is a non-empty, open subset of $\mathcal{S}(B)$, and so there exists $b \in B\setminus\{0\}$ with $\pi(b) \subset a$. Thus $\pi(B\setminus\{0\})$ is a dense subset of $\bar{B}\setminus\{0\}$. Clearly, π is order-preserving, and $a \perp b$ in $B\setminus\{0\}$ if and only if $\pi(a) \perp \pi(b)$ in $\bar{B}\setminus\{0\}$. Thus $(\bar{B},\pi)$ is a completion of $(B\setminus\{0\},<_B)$. ∎

Somewhat abusing notation, we refer to $\bar{B}$ as <u>the completion of B</u>.

2.15 COROLLARY

Let B be a Boolean algebra. Then there is a non-empty set S and a Boolean monomorphism $\pi : B \to \mathcal{P}(S)$.

Proof

Let $S = \mathcal{S}(B)$, and let π be as above. Since $\pi(B) \subset \mathcal{B}(\mathcal{S}(B),\tau_B)$, $\pi(B)$ is a Boolean subalgebra of $\mathcal{P}(S)$. ∎

2.16 EXAMPLE

Let $B = \mathcal{P}(\mathbb{N})$, so that B is a complete Boolean algebra, and the Stone space $S(B)$ consists of the set of ultrafilters on $\mathbb{N}$. Let $\beta\mathbb{N}$ be the character space of the Banach algebra $\ell^\infty(\mathbb{C})$, as in Chapter 1, so that $\beta\mathbb{N}$ is a compact space, and $\mathbb{N}$ is a dense subset of $\beta\mathbb{N}$. We are writing $\bar{\sigma}$ for the closure in $\beta\mathbb{N}$ of a subset σ of $\mathbb{N}$.

Let $\mathfrak{p} \in \beta\mathbb{N}$, and let $\mathcal{U}_{\mathfrak{p}} = \{\sigma \subset \mathbb{N} : \mathfrak{p} \in \bar{\sigma}\}$. Since $\bar{\sigma} \cap \bar{\tau} = \overline{\sigma \cap \tau}$ $(\sigma, \tau \subset \mathbb{N})$, $\sigma \cap \tau \in \mathcal{U}_{\mathfrak{p}}$ if $\sigma, \tau \in \mathcal{U}_{\mathfrak{p}}$. Thus $\mathcal{U}_{\mathfrak{p}}$ is a filter on $\mathbb{N}$. Let $\{\sigma, \tau\}$ be a partition of $\mathbb{N}$. Then $\{\bar{\sigma}, \bar{\tau}\}$ is a partition of $\beta\mathbb{N}$, and so either σ or τ belongs to $\mathcal{U}_{\mathfrak{p}}$. Thus $\mathcal{U}_{\mathfrak{p}}$ is an ultrafilter on $\mathbb{N}$, and we have a map $i : \mathfrak{p} \mapsto \mathcal{U}_{\mathfrak{p}}$, $\beta\mathbb{N} \to S(B)$.

Let $\mathcal{U}$ be an ultrafilter on $\mathbb{N}$, and let $F = \bigcap\{\bar{\sigma} : \sigma \in \mathcal{U}\}$. Since $\{\bar{\sigma} : \sigma \in \mathcal{U}\}$ has the finite intersection property and since $\beta\mathbb{N}$ is compact, F is non-empty. Take $\mathfrak{p} \in F$. Then $\mathcal{U} \subset \mathcal{U}_{\mathfrak{p}}$, and so $\mathcal{U} = \mathcal{U}_{\mathfrak{p}}$. Thus $F = \{\mathfrak{p}\}$, and we have a map $j : \mathcal{U} \mapsto \mathfrak{p}$, $S(B) \to \beta\mathbb{N}$.

Clearly j is the inverse of i, and the map $i : \beta\mathbb{N} \to S(B)$ is a homeomorphism.

Henceforth we identify the Stone space $(S(B), \tau_B)$ of $P(\mathbb{N})$ with the Stone-Čech compactification $\beta\mathbb{N}$ of $\mathbb{N}$.

Points of $\mathbb{N}$ are the <u>fixed</u> (or <u>principal</u>) <u>ultrafilters</u> on $\mathbb{N}$: if $n \in \mathbb{N}$, the corresponding ultrafilter is $\{\sigma \in \mathbb{N} : n \in \sigma\}$. Points of $\beta\mathbb{N} \setminus \mathbb{N}$ are the <u>free ultrafilters</u>. The sets in a free ultrafilter are all infinite.

The set $\beta\mathbb{N} \setminus \mathbb{N}$ is sometimes called the <u>growth</u> of $\mathbb{N}$.

There are even those who believe that $\beta\mathbb{N}$ is defined to be the set of ultrafilters on $\mathbb{N}$ with the Stone topology!

We now turn to the theory of ultraproducts. First, we recall the definition of an ordered field.

2.17 DEFINITION

Let K be a field, and let $<$ be a binary relation on K. Then $(K,<)$ is an <u>ordered field</u> if:

(i) $<$ is a total order on K;

(ii) if $a < b$ in K, then $a + c < b + c$ $(c \in K)$;

(iii) if $a > 0$ and $b > 0$ in K, then $ab > 0$.

Let $(K,<)$ be an ordered field. An element $a \in K$ is <u>positive</u> if $a > 0$.

Set

$$K^+ = \{a \in K : a \geqslant 0\}, \quad K^- = \{a \in K : a \leqslant 0\},$$

$$|a| = \max\{a,-a\} \quad (a \in K).$$

Then $a \in K^+$ if and only if $a = |a|$. We have $|ab| = |a||b|$ and $|a + b| \leqslant |a| + |b|$ $(a,b \in K)$.

Let 1 be the identity of K. The elements p/q (i.e., $(p1)/(q1)$), where $p \in \mathbb{Z}$ and $q \in \mathbb{N}$, constitute a copy of $\mathbb{Q}$ in K, and we identify this subfield with $\mathbb{Q}$.

An ordered field K is <u>real-closed</u> if it has no proper algebraic extension to an ordered field. Equivalently, K is real-closed if, for each $a \in K^+$, there exists $b \in K$ such that $b^2 = a$, and if each polynomial over K of odd degree has a root in K. The canonical example of a real-closed field is $\mathbb{R}$ itself, of course. (The fundamental result on the existence of real-closed fields is the <u>Artin-Schreier theorem</u>: each ordered field K has an algebraic extension to a real-closed field whose order extends the order of K ([41, Theorem 11.4]). We shall not use this theorem.)

Let K be an ordered field. An element $a \in K$ is:

(i) <u>infinitely large</u> if $|a| > n$ $(n \in \mathbb{N})$;

(ii) <u>finite</u> if $|a| < n$ for some $n \in \mathbb{N}$;

(iii) an <u>infinitesimal</u> if $n|a| < 1$ $(n \in \mathbb{N})$.

We denote the set of infinitesimals in K by K^*.

2.18 DEFINITION

Let $\mathcal{U}$ be an ultrafilter on $\mathbb{N}$, and let $f, g \in \mathbb{R}^{\mathbb{N}}$. Set

$$f =_{\mathcal{U}} g \quad \text{if} \quad \{n \in \mathbb{N} : f(n) = g(n)\} \in \mathcal{U},$$

$$f <_{\mathcal{U}} g \quad \text{if} \quad \{n \in \mathbb{N} : f(n) < g(n)\} \in \mathcal{U}.$$

Then $=_{\mathcal{U}}$ is an equivalence relation on $\mathbb{R}^{\mathbb{N}}$; the set of equivalence classes is denoted by $\mathbb{R}^{\mathbb{N}}/\mathcal{U}$, and the equivalence class containing an element f of $\mathbb{R}^{\mathbb{N}}$ is $[f]_{\mathcal{U}}$ or $[f]$. Note that, to define $[f]$, it suffices to specify f on a set of $\mathcal{U}$.

The set $\mathbb{R}^{\mathbb{N}}$ is a real algebra with respect to the pointwise operations on $\mathbb{N}$. For $[f],[g] \in \mathbb{R}^{\mathbb{N}}/\mathcal{U}$ and $\alpha \in \mathbb{R}$, set

$$[f] + [g] = [f + g], \; [f][g] = [fg], \; \alpha[f] = [\alpha f].$$

Clearly, these operations are well-defined on $\mathbb{R}^{\mathbb{N}}/\mathcal{U}$, and $\mathbb{R}^{\mathbb{N}}/\mathcal{U}$ is a real algebra with identify $[1]$. We identify $\alpha \in \mathbb{R}$ with $\alpha[1]$, and so regard $\mathbb{R}$ as a subfield of $\mathbb{R}^{\mathbb{N}}/\mathcal{U}$. Take $[f] \neq [0]$ in $\mathbb{R}^{\mathbb{N}}/\mathcal{U}$, let $\sigma = \{n \in \mathbb{N} : f(n) \neq 0\}$, and set $g(n) = 1/f(n) \quad (n \in \sigma)$. Then $\sigma \in \mathcal{U}$, and $[f][g] = [1]$. Thus $\mathbb{R}^{\mathbb{N}}/\mathcal{U}$ is a field.

Set $[f] < [g]$ (or $[f] <_{\mathcal{U}} [g]$) in $\mathbb{R}^{\mathbb{N}}/\mathcal{U}$ if $f <_{\mathcal{U}} g$. Then $<$ is well-defined on $\mathbb{R}^{\mathbb{N}}/\mathcal{U}$, and it is easily checked that $(\mathbb{R}^{\mathbb{N}}/\mathcal{U},<)$ is an ordered field.

It is also easy to check that $(\mathbb{R}^{\mathbb{N}}/\mathcal{U},<)$ is real-closed: one proves that each positive element is a square and that each polynomial of odd degree over $\mathbb{R}^{\mathbb{N}}/\mathcal{U}$ has a root by working with representatives on $\mathbb{N}$ of the equivalence classes.

Thus, $\mathbb{R}^{\mathbb{N}}/\mathcal{U}$ is a real-closed ordered field.

Let $F : \mathbb{R} \to \mathbb{R}$ be a function. Then F has a natural extension to $\mathbb{R}^{\mathbb{N}}/\mathcal{U}$: define $(F(f))(n) = F(f(n))$ and $F([f]) = [F(f)]$. Again, $F([f])$ is well-defined. More generally, we can consider functions F defined on subsets of $\mathbb{R}^k$, and define $F(a_1,\ldots,a_k)$ for suitable $a_1,\ldots,a_k$ in $\mathbb{R}^{\mathbb{N}}/\mathcal{U}$. For example, the exponentiation function $F(\alpha,\beta) = \beta^{\alpha}$, defined for $\alpha \in \mathbb{R}$ and $\beta > 0$, extends to $F([f],[g]) = [g]^{[f]}$, defined for $[f] \in \mathbb{R}^{\mathbb{N}}/\mathcal{U}$ and $[g] > 0$.

2.19 DEFINITION

Let $\mathcal{U}$ be an ultrafilter on $\mathbb{N}$. Then the ordered field $(\mathbb{R}^{\mathbb{N}}/\mathcal{U}, <_{\mathcal{U}})$ is the <u>ultraproduct</u> of $\mathbb{R}$ by $\mathcal{U}$.

There are also more general notions of ultraproduct, which we shall not require (e.g., [44, §28]).

If $\mathcal{U}$ is a fixed ultrafilter, then $\mathbb{R}^{\mathbb{N}}/\mathcal{U} = \mathbb{R}$. But, if $\mathcal{U}$ is a free ultrafilter, then $\mathbb{R}^{\mathbb{N}}/\mathcal{U}$ is significantly larger than $\mathbb{R}$, and it contains a wealth of infinitely large and of infinitesimal elements. In this case, the ultraproduct $\mathbb{R}^{\mathbb{N}}/\mathcal{U}$ has a natural interpretation as a "non-standard" model of analysis (see [68]): again, we shall not require a discussion of this topic.

Since ℓ^{∞} and c_o are each subalgebras of $\mathbb{R}^{\mathbb{N}}$, we may consider $\ell^{\infty}/\mathcal{U}$ and $c_o/\mathcal{U}$ to be subalgebras of $\mathbb{R}^{\mathbb{N}}/\mathcal{U}$. (If $f \in \ell^{\infty}$ and $g \in \mathbb{R}^{\mathbb{N}}$ with $f =_{\mathcal{U}} g$, then it does not follow that $g \in \ell^{\infty}$, but this fact causes no problems.) Clearly, $\ell^{\infty}/\mathcal{U}$ is the set of finite elements of $\mathbb{R}^{\mathbb{N}}/\mathcal{U}$, and $c_o/\mathcal{U}$ is an ideal in the algebra $\ell^{\infty}/\mathcal{U}$. Note that, if $a \in c_o/\mathcal{U}$ and if $b \in \mathbb{R}^{\mathbb{N}}/\mathcal{U}$ with $|b| \leq |a|$, then $b \in c_o/\mathcal{U}$.

We can now give an alternative description of some algebras which arose in Chapter 1. Let $p \in \beta\mathbb{N}$, and let $\mathcal{U}$ be the corresponding ultrafilter on $\mathbb{N}$. We now write M_p and J_p for ideals in the algebra $C(\beta\mathbb{N}) = \ell^{\infty}$ of *real*-valued, continuous functions on $\beta\mathbb{N}$. Take $f \in C(\beta\mathbb{N})$. Then $f \in J_p$ if and only if $f|\bar{\sigma} = 0$ for some $\sigma \in \mathcal{U}$. Thus

$f \in J_p$ if and only if $f =_{\mathcal{U}} 0$, and so $\ell^\infty/\mathcal{U}$ is isomorphic (as a real algebra) to $C(\beta\mathbb{N})/J_p$; we identify these two algebras. The quotient M_p/J_p corresponds to $(\mathbb{R}^{\mathbb{N}}/\mathcal{U})^*$, the set of infinitesimals of $\mathbb{R}^{\mathbb{N}}/\mathcal{U}$, and we write $(\ell^\infty/\mathcal{U})^*$ for this algebra. It is an integral domain whose quotient field is $\mathbb{R}^{\mathbb{N}}/\mathcal{U}$. The quotient $c_o/(J_p \cap c_o)$ corresponds to $c_o/\mathcal{U}$.

We reformulate the reduction theorem, Theorem 1.8, of Chapter 1. Clearly, we can suppose that the homomorphism θ of Theorem 1.8 does not vanish on the real subalgebra c_o of $c_o(\mathbb{C})$. Thus we have the following two results.

2.20 THEOREM

Let X be a compact space. Assume that there is a discontinuous homomorphism from $C(X,\mathbb{C})$ into a Banach algebra. Then there is a free ultrafilter $\mathcal{U}$ on $\mathbb{N}$, a radical Banach algebra R, and a non-zero homomorphism from $c_o/\mathcal{U}$ into R. ∎

2.21 THEOREM

There is a discontinuous homomorphism from $\ell^\infty(\mathbb{C})$ (respectively, from $c_o(\mathbb{C})$) into a Banach algebra if and only if there is a free ultrafilter $\mathcal{U}$ on $\mathbb{N}$ such that $(\ell^\infty/\mathcal{U})^*$ (respectively, $c_o/\mathcal{U}$) is seminormable. ∎

As a final topic in this chapter, we discuss P-points in the space $\beta\mathbb{N}\setminus\mathbb{N}$; Theorem 2.24 will be used in Proposition 3.7.

2.22 DEFINITION

Let X be a topological space. A point $x \in X$ is a <u>P-point</u> if each G_δ-set in X containing x is a neighbourhood of x.

Let x belong to a compact space X. Suppose first that x is a P-point, and that $f \in M_x$. Since $Z(f)$ is a G_δ-set in X, $f \in J_x$, and so $M_x = J_x$. Suppose secondly that $M_x = J_x$, and that V is a G_δ-set containing x, say $V = \cap V_n$, where $V_n \in N_x$. Take $f_n \in M_x$ with $0 \leqslant f_n(y) \leqslant 1$ $(y \in X)$ and $f_n(X \setminus V_n) = \{1\}$, and set $f = \Sigma 2^{-n} f_n$. Then $f \in M_x$ and $Z(f) \subset V$. Since $Z(f) \in N_x$, x is a P-point. Thus x is a P-point of X if and only if $M_x = J_x$.

The only P-points of $\beta\mathbb{N}$ are the points of $\mathbb{N}$. We shall be concerned with points $\mathfrak{p}$ which are P-points in the compact space $\beta\mathbb{N} \setminus \mathbb{N}$.

Clearly, each infinite compact space must contain points which are not P-points, and so $\beta\mathbb{N} \setminus \mathbb{N}$ contains non-P-points. It is a result of W. Rudin (see [69, 4.30 and 4.31]) that, if CH holds, then $\beta\mathbb{N} \setminus \mathbb{N}$ contains a dense set of $2^{\mathfrak{c}}$ P-points and a dense set of $2^{\mathfrak{c}}$ non-P-points (where $\mathfrak{c} = 2^{\aleph_0}$).

2.23 DEFINITION

Let $\mathcal{U}$ be a free ultrafilter on $\mathbb{N}$, and let $\{\sigma_k : k \in \mathbb{N}\}$ be a countable partition of $\mathbb{N}$. Then $\mathcal{U}$ is <u>quasi-selective</u> for $\{\sigma_k\}$ if there exists $\sigma \in \mathcal{U}$ such that $\sigma \cap \sigma_k$ is finite for each $k \in \mathbb{N}$.

2.24 THEOREM

Let $\mathfrak{p} \in \beta\mathbb{N} \setminus \mathbb{N}$, and let $\mathcal{U}$ be the corresponding ultrafilter. Then the following conditions are equivalent:

(a) $\mathfrak{p}$ is a P-point of $\beta\mathbb{N} \setminus \mathbb{N}$;

(b) $M_{\mathfrak{p}}/J_{\mathfrak{p}} = c_0/(J_{\mathfrak{p}} \cap c_0)$;

(c) $(\ell^\infty/\mathcal{U})^* = c_0/\mathcal{U}$;

(d) $\mathcal{U}$ is quasi-selective for each partition $\{\sigma_k : k \in \mathbb{N}\}$ of $\mathbb{N}$ for which $\sigma_k \notin \mathcal{U}$ $(k \in \mathbb{N})$.

<u>Proof</u>

Conditions (b) and (c) are reformulations of each other.

(a) $\Rightarrow$ (b). Take $f \in M_{\mathfrak{p}}$. By (a), there is a clopen neighbourhood U of p in $\beta\mathbb{N}$ such that $f|U \cap (\beta\mathbb{N}\setminus\mathbb{N}) = 0$. Set $g(x) = f(x)$ $(x \in U)$, $g(x) = 0$ $(x \in \beta\mathbb{N}\setminus U)$. Then $g \in c_o$ and $f - g \in J_p$. Thus (b) holds.

(b) $\Rightarrow$ (a). This is similarly easy.

(a) $\Rightarrow$ (d). Let $\{\sigma_k : k \in \mathbb{N}\}$ be a partition of $\mathbb{N}$ such that $\sigma_k \notin \mathcal{U}$ $(k \in \mathbb{N})$. Let $V_k = (\beta\mathbb{N}\setminus\mathbb{N})\setminus\bar{\sigma}_k$ $(k \in \mathbb{N})$, and let $V = \bigcap V_k$. Then V is a G_δ-set in $\beta\mathbb{N}\setminus\mathbb{N}$ containing p, and so, by (a), V is a neighbourhood of p in $\beta\mathbb{N}\setminus\mathbb{N}$. Take $\sigma \subset \mathbb{N}$ such that $p \in \bar{\sigma} \subset V$. Then $\sigma \in \mathcal{U}$. For $k \in \mathbb{N}$, $\overline{\sigma \cap \sigma_k} = \bar{\sigma} \cap \bar{\sigma}_k \subset \mathbb{N}$, and so $\sigma \cap \sigma_k$ is finite. Thus (d) holds.

(d) $\Rightarrow$ (a). Let V be a G_δ-set in $\beta\mathbb{N}\setminus\mathbb{N}$ containing p. Then there exists $(\tau_k) \subset P(\mathbb{N})$ such that $\tau_{k+1} \subsetneq \tau_k$ and $p \in \bigcap\bar{\tau}_k \subset V \cup \mathbb{N}$. Set

$$\sigma_1 = \mathbb{N}\setminus\tau_1, \quad \sigma_{k+1} = \tau_k\setminus\tau_{k+1} \quad (k \in \mathbb{N}).$$

Then $\{\sigma_k : k \in \mathbb{N}\}$ is a partition of $\mathbb{N}$, and $\sigma_k \notin \mathcal{U}$ $(k \in \mathbb{N})$. By (d), there exists $\sigma \in \mathcal{U}$ such that $\sigma \cap \sigma_k$ is finite for each $k \in \mathbb{N}$. Then $p \in \bar{\sigma}$ and $\bar{\sigma}\setminus\mathbb{N} \subset \bar{\tau}_k\setminus\mathbb{N}$ $(k \in \mathbb{N})$, so that $\bar{\sigma}\setminus\mathbb{N} \subset V$. Thus V is a neighbourhood of p in $\beta\mathbb{N}\setminus\mathbb{N}$, and (a) holds. ∎

2.25 NOTES

The discussion of partially ordered sets and of Boolean algebras is standard. For an introduction to Boolean algebras, see [39].

Proposition 2.4 is taken from [72].

Theorem 2.10 is given explicitly in [49, II, Lemma 3.3]. Let (B,π) be the completion of P constructed in 2.10. It can be shown that the following conditions are equivalent: (a) π is an embedding; (b) U_a is regular-open in (P,τ) for each $a \in P$; (c) P is separative.

The completion of a Boolean algebra is constructed in [14] and [69], for example. It is shown in [14, 2.33] that the following conditions on a Boolean algebra B are equivalent: (a) B is complete; (b) the map $\pi : B \to \bar{B}$ of 2.14 is a surjection; (c) the space $(S(B), \tau_B)$ is extremely disconnected.

The theory of ultraproducts is given in [7], [9], and [44], for example, and P-points are discussed in [9], [36], and [69]. Theorem 2.24 is well known (see [69, Exercise 4A]). It is a theorem of Shelah ([61, VI, §4], [71]) that there are models of set theory (see Chapter 4) in which the space $\beta\mathbb{N}\backslash\mathbb{N}$ contains no P-points. Thus the existence of P-points in $\beta\mathbb{N}\backslash\mathbb{N}$ is independent of the theory ZFC.

3 WOODIN'S CONDITION

We now give the key set-theoretic consequence of the existence of an incomplete norm on an algebra $C(X,\mathbb{C})$. It is this condition which will be applied (in Theorem 6.25) as a step in the proof that there are models of set theory in which there are no discontinuous homomorphisms from any algebra $C(X,\mathbb{C})$, and hence in which all norms on each $C(X,\mathbb{C})$ are equivalent to the uniform norm.

In the second part of this chapter, we shall discuss the question, left open in Chapter 1, whether or not the existence of a discontinuous homomorphism from $c_o(\mathbb{C})$ entails the existence of such a map from $\ell^\infty(\mathbb{C})$.

Recall first from Examples 2.2 that we write $<<_F$ for the strong Fréchet order on $\mathbb{N}^{\mathbb{N}}$, and $<<$ for the divisibility order on $A\backslash\{0\}$, where A is a radical algebra. With a slight abuse of notation, we write $(A,<<)$ for $(A\backslash\{0\},<<)$.

3.1 DEFINITION

Let $g \in \mathbb{N}^{\mathbb{N}}$ with $1 <<_F g$. Then

$$\langle g\rangle = \{f \in \mathbb{N}^{\mathbb{N}} : 1 <<_F f <<_F g\},$$

where 1 denotes the constant function $n \mapsto 1$ on $\mathbb{N}$.

Thus $\langle g\rangle$ is a subset of the set of functions whose graph is "eventually under" the graph of g. Certainly $(\langle g\rangle,<<_F)$ is a partially ordered set. Let $\mathcal{U}$ be an ultra-filter on $\mathbb{N}$. Then we regard $(\langle g\rangle/\mathcal{U},<_{\mathcal{U}})$ as a totally ordered subset of the ultraproduct $(\mathbb{R}^{\mathbb{N}}/\mathcal{U},<_{\mathcal{U}})$.

3.2 THEOREM

Let X be a compact space. Assume that there is a discontinuous homomorphism from $C(X,\mathbb{C})$ into a Banach algebra. Then there is an unbounded, monotone-increasing function g in $\mathbb{N}^{\mathbb{N}}$, a free ultrafilter $\mathcal{U}$ on $\mathbb{N}$, and an embedding of $(\langle g\rangle/\mathcal{U},<_{\mathcal{U}})$ into $(\mathbb{N}^{\mathbb{N}},<_F)$.

Proof

By 2.20, there is a free ultrafilter $\mathcal{U}$ on $\mathbb{N}$, a radical Banach algebra R, and a non-zero homomorphism $\theta : c_o/\mathcal{U} \to R$. We write $<$ for $<_{\mathcal{U}}$.

First, we define g. Choose $a \in c_o/\mathcal{U}$ with $\theta(a) \neq 0$; we may suppose that $a > 0$. Then choose $\alpha = (\alpha_n) \in c_o$ with $[\alpha] = a$; by redefining α off a set in $\mathcal{U}$, we may suppose that $0 < \alpha_n < 1$ $(n \in \mathbb{N})$. Now choose g to be an unbounded, monotone-increasing function in $\mathbb{N}^{\mathbb{N}}$ such that

$$\frac{1}{g(n)^2}\log\left(\frac{1}{\alpha_n}\right) \to \infty \quad \text{as} \quad n \to \infty. \tag{1}$$

For convenience, we also choose g so that $g(n) \geq 2$ $(n \in \mathbb{N})$.

We now construct an embedding $\pi : (\langle g\rangle/\mathcal{U},<) \to (\mathbb{N}^{\mathbb{N}},<_F)$ as a composition of maps:

$$(\langle g\rangle/\mathcal{U},<) \xrightarrow{\nu} (c_o/\mathcal{U},<<) \xrightarrow{\theta} (R\backslash \mathrm{nil}\, R,<<) \xrightarrow{\tau} (\mathbb{N}^{\mathbb{N}},<_F).$$

The map θ is given (as a map into R), and clearly θ is isotonic. The maps ν and τ will be anti-isotonic.

Take $f \in \langle g\rangle$, and set

$$\alpha^{f/g^2} = (\alpha_n^{f(n)/g(n)^2}).$$

Since $f(n) \geq 1$ $(n \in \mathbb{N})$, it follows from (1) that $\alpha^{f/g^2} \in c_o\backslash\{0\}$. Now take $[f] \in \langle g\rangle/\mathcal{U}$, and set

$$\nu([f]) = [\alpha^{f/g^2}].$$

Then ν is well-defined. Take $[f_1] < [f_2]$ in $\langle g\rangle/\mathcal{U}$. By redefining f_1 and f_2 off a set in $\mathcal{U}$, we may suppose that $f_1(n) < f_2(n) \leqslant g(n)$ $(n \in \mathbb{N})$. Set

$$\beta_n = \alpha_n^{(f_2(n) - f_1(n))/g(n)^2} \qquad (n \in \mathbb{N}).$$

Again by (1), $(\beta_n) \in c_o$. Also, $\nu([f_1])[(\beta_n)] = \nu([f_2])$, and so $\nu([f_2]) \ll \nu([f_1])$ in $c_o/\mathcal{U}$. Thus

$$\nu : (\langle g\rangle/\mathcal{U},<) \to (c_o/\mathcal{U},\ll)$$

is an anti-isotonic map.

We next *claim* that the range of $\theta\circ\nu$ is contained in $R\backslash\text{nil } R$. For suppose that $[f] \in \langle g\rangle/\mathcal{U}$. Take $k \in \mathbb{N}$, and set

$$\beta_n = \alpha_n^{1 - (kf(n)/g(n)^2)} \qquad (n \in \mathbb{N}).$$

Then $1 - (kf(n)/g(n)^2) > 1/2$ eventually, and so $\beta_n < \alpha_n^{1/2}$ eventually. Thus $\beta = (\beta_n) \in c_o$. Also,

$$\beta_n \alpha_n^{kf(n)/g(n)^2} = \alpha_n \qquad (n \in \mathbb{N}),$$

so that $(\nu([f]))^k[\beta] = a$, and hence $((\theta\circ\nu)([f]))^k \neq 0$, establishing the claim.

Let $\|\cdot\|$ be the norm on R. For $x \in R\backslash\text{nil } R$, define

$$\tau(x)(n) = \min\{k \in \mathbb{N} : k \geqslant \|x^n\|^{-1}\} \qquad (n \in \mathbb{N}).$$

Then $\tau(x) \in \mathbb{N}^{\mathbb{N}}$. Take $x,y \in R\backslash\text{nil } R$ with $x \ll y$, say $x = yz$, where $z \in R$. Then $z \notin \text{nil } R$, and

$$\|x^n\|^{-1} \geqslant \|y^n\|^{-1}\|z^n\|^{-1} \qquad (n \in \mathbb{N}).$$

Since R is a radical Banach algebra, $\|y^n\|^{-1} \to \infty$ and $\|z^n\|^{-1} \to \infty$ as $n \to \infty$, and so $\tau(x)(n) > \tau(y)(n)$

eventually. Thus $\tau(y) <_F \tau(x)$, and so

$$\tau : (R\backslash \text{nil } R, <<) \to (\mathbb{N}^{\mathbb{N}}, <_F)$$

is an anti-isotonic map.

Let $\pi = \tau \circ \theta \circ \nu$. Since θ is isotonic and both ν and τ are anti-isotonic, π is isotonic. Since $<$ is a total order on $\langle g\rangle/\mathcal{U}$, τ is an embedding. ∎

The following condition, <u>Woodin's condition</u>, gives the set theoretic consequence of the existence of discontinuous homomorphisms that we shall apply.

Define $Z(n) = n \quad (n \in \mathbb{N})$, so that $Z \in \mathbb{N}^{\mathbb{N}}$.

3.3 THEOREM

Let X be a compact space. Assume that there is a discontinuous homomorphism from $C(X,\mathbb{C})$ into a Banach algebra. Then there is a free ultrafilter $\mathcal{V}$ on $\mathbb{N}$ and an embedding from $(\langle Z\rangle/\mathcal{V}, <_{\mathcal{V}})$ into $(\mathbb{N}^{\mathbb{N}}, <_F)$.

<u>Proof</u>

Let g be the function and $\mathcal{U}$ the free ultrafilter specified in Theorem 3.2. Set $\sigma_n = g^{-1}(\{n\})$, so that $\{\sigma_n : n \in \mathbb{N}\}$ is a partition of $\mathbb{N}$, and set

$$\mathcal{V} = \{\sigma \subset \mathbb{N} : \cup\{\sigma_n : n \in \sigma\} \in \mathcal{U}\}.$$

Certainly, $\mathcal{V}$ is a filter on $\mathbb{N}$. If $\sigma \subset \mathbb{N}$, then $\sigma \in \mathcal{V}$ or $\mathbb{N}\backslash\sigma \in \mathcal{V}$, and so $\mathcal{V}$ is an ultrafilter on $\mathbb{N}$. Since $g(n) \to \infty$ as $n \to \infty$, each σ_n is finite, and so $\mathcal{V}$ cannot be a fixed ultrafilter on $\mathbb{N}$.

Take $f \in \langle Z\rangle/\mathcal{V}$, so that $f(n) \to \infty$ and $n - f(n) \to \infty$ as $n \to \infty$. Then $(f\circ g)(n) \to \infty$ and $g(n) - (f\circ g)(n) \to \infty$ as $n \to \infty$, and so $f\circ g \in \langle g\rangle$. The map

$$[f]_{\mathcal{V}} \mapsto [f\circ g]_{\mathcal{U}}, \quad (\langle Z\rangle/\mathcal{V}, <_{\mathcal{V}}) \to (\langle g\rangle/\mathcal{U}, <_{\mathcal{U}}),$$

is well defined and isotonic. Thus the result follows from 3.2. ∎

The force of the above theorem is, of course, that, if we can show that in certain circumstances there is no embedding from $(\langle Z\rangle/V, <_V)$ into $(\mathbb{N}^{\mathbb{N}}, <_F)$ for any free ultrafilter V on $\mathbb{N}$ (as we shall in Chapters 6 and 8), then it will follow that each homomorphism from each algebra $C(X,\mathbb{C})$ into a Banach algebra is automatically continuous.

We now discuss the relation between the existence of discontinuous homomorphisms from $\ell^\infty(\mathbb{C})$ and from $c_o(\mathbb{C})$. In this discussion, U and V denote free ultrafilters on $\mathbb{N}$. The fact that we do know is the following, an immediate consequence of Theorems 2.20 and 2.21.

Assume that there exists U *such that* $(\ell^\infty/U)^*$ *is seminormable. Then there exists* V *such that* c_o/V *is seminormable.*

This suggests the following questions.

(1) *Assume that there exists* U *such that* c_o/U *is seminormable. Does there exist* V *such that* $(\ell^\infty/V)^*$ *is seminormable?*

(2) *Assume that* c_o/U *is seminormable. Is* $(\ell^\infty/U)^*$ *seminormable?*

(3) *Assume that* $(\ell^\infty/U)^*$ *is seminormable. Is* c_o/U *seminormable?*

Question (1) is equivalent to our main question: if there is a discontinuous homomorphism from $C(X,\mathbb{C})$ for some compact space X, is there a discontinuous homomorphism from $C(Y,\mathbb{C})$ for every infinite compact space Y? We do not know the answer to this question. Question (2) is a variant of (1): we cannot solve this question either. Conceivably, it is a theorem of ZFC that, for each free ultrafilter U

and each $a \in c_o/U$, there is a monomorphism $(\ell^\infty/U)^* \to c_o/U$ whose range contains a. If so, we would have a positive answer to Question (2), and hence to Question (1). This seems to be an interesting problem to attack, independent of its applications to the above questions.

Although we cannot solve Question (3) either, we do have a partial result, which we now discuss.

Thus, let U be a free ultrafilter on $\mathbb{N}$, corresponding to the point $\mathfrak{p} \in \beta\mathbb{N} \setminus \mathbb{N}$, say, and assume that $(\ell^\infty/U)^*$ is seminormable. If $\mathfrak{p}$ is a P-point of $\beta\mathbb{N} \setminus \mathbb{N}$, then, by Theorem 2.24, $(\ell^\infty/U)^* = c_o/U$, and so, trivially, c_o/U is seminormable. Now suppose that $\mathfrak{p}$ is not a P-point of $\beta\mathbb{N} \setminus \mathbb{N}$, and take $a \in (M_\mathfrak{p}/J_\mathfrak{p}) \setminus (c_o/J_\mathfrak{p})$, say $a = f + J_\mathfrak{p}$. If $a^n \in c_o/J_\mathfrak{p}$, then there is a clopen neighbourhood U of $\mathfrak{p}$ in $\beta\mathbb{N}$ such that $f^n = 0$ on $U \cap (\beta\mathbb{N} \setminus \mathbb{N})$. Set $g(x) = f(x)$ $(x \in U)$, $g(x) = 0$ $(x \in \beta\mathbb{N} \setminus U)$. Then $g \in c_o$ and $f - g \in J_\mathfrak{p}$, a contradiction. Hence $a^n \notin c_o/J_\mathfrak{p}$ $(n \in \mathbb{N})$, and so there is a prime ideal P in $M_\mathfrak{p}$ with $c_o/J_\mathfrak{p} \subset P/J_\mathfrak{p}$ and $a \notin P/J_\mathfrak{p}$. By 1.11(i), the Continuum Hypothesis implies that there is a homomorphism θ from ℓ^∞ into a Banach algebra such that $\ker \theta = P$. The homomorphism θ gives a non-zero seminorm on $(\ell^\infty/U)^*$, but its restriction to c_o/U is zero. Thus, if $\mathfrak{p}$ is not a P-point of $\beta\mathbb{N} \setminus \mathbb{N}$, there is no trivial positive solution to Question (3).

Assume that there is a discontinuous homomorphism from $\ell^\infty(\mathbb{C})$. The following result shows that, if the answer to Question (3) is positive, then c_o/V is seminormable for each free ultrafilter V.

3.4 PROPOSITION

Assume that there exist free ultrafilters U and V such that $(\ell^\infty/U)^*$ is seminormable and c_o/V is not seminormable. Then there is a free ultrafilter W such that $(\ell^\infty/W)^*$ is seminormable, but c_o/W is not seminormable.

<u>Proof</u>

Let $\{\sigma_k : k \in \mathbb{N}\}$ be a partition of $\mathbb{N}$ into infinite subsets, let $h_k : \sigma_k \to \mathbb{N}$ be a bijection, and let $H_k : \mathcal{P}(\sigma_k) \to \mathcal{P}(\mathbb{N})$ be the induced bijection. For $k \in \mathbb{N}$, let $\mathcal{V}_k = \{\sigma \subset \mathbb{N} : H_k(\sigma \cap \sigma_k) \in \mathcal{V}\}$. Then each $\mathcal{V}_k$ is a free ultrafilter on $\mathbb{N}$; the corresponding point in $\beta\mathbb{N} \setminus \mathbb{N}$ is p_k, say. Define

$$\mathcal{W} = \{\sigma \subset \mathbb{N} : \{k \in \mathbb{N} : \sigma \in \mathcal{V}_k\} \in \mathcal{U}\}.$$

Then $\mathcal{W}$ is also a free ultrafilter on $\mathbb{N}$.

For $f \in \ell^\infty$, set $\rho(f)(k) = f(p_k)$. Then $\rho : \ell^\infty \to \ell^\infty$ is a homomorphism. Take $f_1, f_2 \in \ell^\infty$, and let $\tau = \{m : f_1(m) = f_2(m)\}$. If $\tau \in \mathcal{V}_k$, then $f_1(p_k) = f_2(p_k)$, and so $\rho(f_1)(k) = \rho(f_2)(k)$. So, if $f_1 =_{\mathcal{W}} f_2$, then $\tau \in \mathcal{W}$, $\{k : \rho(f_1)(k) = \rho(f_2)(k)\} \in \mathcal{U}$, and $\rho(f_1) =_{\mathcal{U}} \rho(f_2)$. Thus ρ induces a homomorphism $\bar{\rho} : \ell^\infty/\mathcal{W} \to \ell^\infty/\mathcal{U}$. Take $g \in \ell^\infty$, and set $f(n) = g(k)$ if $n \in \sigma_k$. Then $f \in \ell^\infty$ and $\rho(f) = g$. Thus ρ, and hence $\bar{\rho}$, are surjections. Since $(\ell^\infty/\mathcal{U})^*$ is seminormable, $(\ell^\infty/\mathcal{W})^*$ is seminormable.

For $f \in \ell^\infty$ and $n \in \mathbb{N}$, take $k \in \mathbb{N}$ such that $n \in \sigma_k$, and set

$$\mu(f)(n) = (f \circ h_k)(n), \quad \tilde{\mu}(f)(n) = \min\{\mu(f)(n), 1/k\}.$$

Then $\mu : \ell^\infty \to \ell^\infty$ is a homomorphism. Clearly, if $f_1 =_{\mathcal{V}} f_2$, then $\mu(f_1) =_{\mathcal{W}} \mu(f_2)$, and so μ induces a homomorphism $\bar{\mu} : \ell^\infty/\mathcal{V} \to \ell^\infty/\mathcal{W}$. Take $f \in c_o$: we see that $\tilde{\mu}(f) \in c_o$. Set

$$\tau = \{m : \mu(f)(m) = \tilde{\mu}(f)(m)\}.$$

For each k, $\sigma_k \setminus (\tau \cap \sigma_k)$ is finite, and so $\tau \in \mathcal{V}_k$. Thus $\tau \in \mathcal{W}$. Also, $[\mu(f)]_{\mathcal{W}} = [\tilde{\mu}(f)]_{\mathcal{W}}$, and so $[\mu(f)]_{\mathcal{W}} \in c_o/\mathcal{W}$ (although, in general, $\mu(f) \notin c_o$). Hence $\bar{\mu}(c_o/\mathcal{V}) \subset c_o/\mathcal{W}$.

Suppose that p is a non-zero seminorm on $c_o/\mathcal{W}$, and choose $b \in c_o/\mathcal{W}$ with $p(b) \neq 0$. Take $g \in c_o$ with $[g]_{\mathcal{W}} = |b|^{1/2}$, and let $f_k = (g|\sigma_k) \circ h_k^{-1}$ $(k \in \mathbb{N})$. Then

$(f_k) \subset c_o$. Choose $f \in c_o$ such that, for each k, $|f_k(n)| < f(n)$ eventually, and set $a = [f]_V$, so that $\bar{\mu}(a) = [\mu(f)]_W$ by definition. Thus

$$|b| = [g^2]_W <_W [\mu(f^2)]_W = \bar{\mu}(a)^2,$$

and so $b << \bar{\mu}(a)$ in c_o/W. Let $q = p\circ\bar{\mu}$. Then q is a seminorm on c_o/V, and q is non-zero because $q(a) \neq 0$. But this is a contradiction, because c_o/V is not seminormable. Hence c_o/W is not seminormable. ∎

We shall explain in Chapter 6 that, if the Continuum Hypothesis holds, then, for any two free ultrafilters U and V, the four algebras $(\ell^\infty/U)^*$, $(\ell^\infty/V)^*$, c_o/U, and c_o/V are all isomorphic (as real algebras). Should the Continuum Hypothesis fail, however, then $(\ell^\infty/U)^*$ and $(\ell^\infty/V)^*$ differ radically whenever $U \neq V$.

In Theorem 3.2, we only made use of the order structure (with respect to divisibility) of the radical Banach algebra R, and we essentially ignored the structure of R as an algebra. It may be that we are losing information by doing this. This idea motivates us to make the following definition.

3.5 DEFINITION

Let A be an algebra. Then A is weakly seminormable if there is a non-empty subset S of $A\backslash\{0\}$ such that:

(i) if $b \in S$ and $b << a$, then $a \in S$;

(ii) there is an isotonic map from $(S,<<)$ into $(\mathbb{N}^{\mathbb{N}},<_F)$.

It follows from 3.2 and 2.4 that a radical Banach algebra is weakly seminormable (with $S = R\backslash \text{nil } R$).

The next result shows that the set-theoretic condition we are going to use cannot distinguish between the hypothesis that c_o/U is seminormable and that it is weakly seminormable.

3.6 PROPOSITION

Assume that there is a free ultrafilter $\mathcal{U}$ on $\mathbb{N}$ such that $c_o/\mathcal{U}$ is weakly seminormable. Then there is a free ultrafilter $\mathcal{V}$ on $\mathbb{N}$ and an embedding from $(\langle Z\rangle/\mathcal{V}, <_{\mathcal{V}})$ into $(\mathbb{N}^{\mathbb{N}}, <_F)$.

Proof

Let $S \subset c_o/\mathcal{U}$ be as specified in 3.5, and take $b \in S$. Set $a = |b|^{1/2}$. Then $b << a$, and so $a \in S$.

Choose $g \in \mathbb{N}^{\mathbb{N}}$ as in 3.2, and let

$$\nu : (\langle g\rangle/\mathcal{U}, <_{\mathcal{U}}) \to (c_o/\mathcal{U}, <<)$$

be the anti-isotonic map constructed in the proof of 3.2. If $[f] \in \langle g\rangle/\mathcal{U}$, then $a << \nu([f])$ in $c_o/\mathcal{U}$, and so $\nu([f]) \in S$. Thus the range of ν is contained in S.

By hypothesis, there is an isotonic map from $(S,<<)$ into $(\mathbb{N}^{\mathbb{N}},<_F)$, and so, by 2.4, there is an anti-isotonic map from $(S,<<)$ into $(\mathbb{N}^{\mathbb{N}},<_F)$.

The result now follows as in the proof of 3.3. ∎

The following result shows that the analogue of Question (1), above, with "seminormable" replaced by "weakly seminormable" does have a positive answer.

3.7 PROPOSITION

Assume that there is a free ultrafilter $\mathcal{U}$ on $\mathbb{N}$ such that $c_o/\mathcal{U}$ is weakly seminormable. Then there is a free ultrafilter $\mathcal{V}$ on $\mathbb{N}$ such that $(\ell^\infty/\mathcal{V})^*$ is weakly seminormable.

Proof

Let $\mathcal{U}$ correspond to the point p of $\beta\mathbb{N}\setminus\mathbb{N}$. We may suppose that p is not a P-point of $\beta\mathbb{N}\setminus\mathbb{N}$, for otherwise, by 2.24, $(\ell^\infty/\mathcal{U})^* = c_o/\mathcal{U}$ and the result is immediate.

By Theorem 2.24, there is a partition $\{\sigma_k : k \in \mathbb{N}\}$ of $\mathbb{N}$ such that $\sigma_k \notin \mathcal{U}$ $(k \in \mathbb{N})$ and such that $\mathcal{U}$ is not quasi-selective for $\{\sigma_k\}$. Set

$$V = \{\sigma \subset \mathbb{N} : \cup\{\sigma_k : k \in \sigma\} \in U\}.$$

As in Theorem 3.3, V is an ultrafilter on $\mathbb{N}$. Since $\sigma_k \notin U$ $(k \in \mathbb{N})$, V is a free ultrafilter.

We now make a *remark*. Let $\alpha = (\alpha_n) \in c_o$ with $\alpha_n > 0$, and let $\beta = (\beta_n) \in \ell^\infty$ be such that $\beta_n > 0$ and β is constant on each set σ_k. Then $[\alpha^\beta]_U \in c_o/U$. For, given $n \in \mathbb{N}$, take $k \in \mathbb{N}$ such that $n \in \sigma_k$, and set

$$\gamma_n = \min\{\alpha_n^{\beta_n}, 1/k\}, \quad \gamma = (\gamma_n).$$

Then $\gamma \in c_o$: for each $\varepsilon > 0$, $\{n : |\gamma| > \varepsilon\}$ is contained in the union of finitely many of the sets σ_k, and it contains only finitely many points of each σ_k because $\alpha \in c_o$ and β is a positive constant on each σ_k. Let $\sigma = \{n : \gamma_n \neq \alpha_n^{\beta_n}\}$. Then $\sigma \cap \sigma_k$ is finite for each k, and so, since U is not quasi-selective for $\{\sigma_k\}$, $\sigma \notin U$. Hence $\mathbb{N}\setminus\sigma \in U$, and $\gamma =_U \alpha^\beta$, proving the remark.

Let $S \subset c_o/U$ be as specified in 3.5, and take $\alpha = (\alpha_n) \in c_o$ such that $[\alpha]_U \in S$. By replacing α by $|\alpha|^{1/2}$, we may suppose that $\alpha_n > 0$ $(n \in \mathbb{N})$. For $f \in \ell^\infty$, set $\beta(f)(n) = |f(k)|$ for $n \in \sigma_k$ if $f(k) \neq 0$, and set $\beta(f)(n) = 1/2$ for $n \in \sigma_k$ if $f(k) = 0$. Set

$$\rho(f) = \alpha^{\beta(f)} \quad (f \in \ell^\infty).$$

By the above remark, $[\rho(f)]_U \in c_o/U$ $(f \in \ell^\infty)$. Clearly, if $f_1 =_V f_2$, then $\rho(f_1) =_U \rho(f_2)$, and so ρ induces a map $\bar{\rho} : \ell^\infty/V \to c_o/U$.

Take $a \in (\ell^\infty/V)^*\setminus\{0\}$, say $a = [f]_V$. We may suppose that $|f(n)| < 1$ $(n \in \mathbb{N})$. Since

$$(1 - \beta(f))(n) > 0 \quad (n \in \mathbb{N}),$$

the above remark shows that $\left[\alpha^{1-\beta(f)}\right]_U \in c_o/U$, and so

$[\alpha]_{\mathcal{U}} << \bar{\rho}(a)$ in $c_o/\mathcal{U}$. Thus $\bar{\rho}(a) \in S$.

Take $a_1, a_2 \in (\ell^\infty/\mathcal{V})^* \backslash \{0\}$ with $a_1 << a_2$, say $a_1 = [f_1]_{\mathcal{V}}$, $a_2 = [f_2]_{\mathcal{V}}$. We may suppose that

$$0 < |f_1(n)| < |f_2(n)| \quad (n \in \mathbb{N}),$$

and the above remark then shows that

$$\left[\alpha^{\beta(f_2) - \beta(f_1)}\right]_{\mathcal{U}} \in c_o/\mathcal{U}.$$

It follows that $\bar{\rho}(f_2) << \bar{\rho}(f_1)$ in $c_o/\mathcal{U}$, and that $\bar{\rho} : ((\ell^\infty/\mathcal{V})^*, <<) \to (c_o/\mathcal{U}, <<)$ is anti-isotonic. By hypothesis, there is an isotonic map from $(S, <<)$ into $(\mathbb{N}^{\mathbb{N}}, <_{\mathcal{F}})$. The result follows, again using 2.4. ∎

The above result suggests a final question. Let $\mathcal{V}$ be a free ultrafilter on $\mathbb{N}$.

If $(\ell^\infty/\mathcal{V})^*$ *is weakly seminormable, is* $(\ell^\infty/\mathcal{V})^*$ *seminormable?*

We are asking if an order-theoretic condition on $(\ell^\infty/\mathcal{V})^*$ implies an apparently stronger algebraic condition. If this were the case, then, again, we would have a positive answer to our main question, Question (1).

3.8 NOTES

Our version of Woodin's condition is a reformulation of the one given in [72]. As we stated in the Introduction, Solovay first used the condition to show that there is a model of set theory in which no map specified by the condition exists, and hence that there is a model in which all homomorphisms from each $C(X,\mathbb{C})$ are continuous.

In the proof of Theorem 3.2, we showed that the range of $\theta \circ \nu$ is contained in $R\backslash \text{nil } R$. A more general result is due, independently, to Esterle and to Sinclair: if $a \in c_o/\mathcal{U}$ and $\theta(a) \in \text{nil } R$, then $\theta(a) = 0$ (see [65, 11.7] and [17]).

Proposition 3.4 is new, but 3.7 is from [72].

4 INDEPENDENCE IN SET THEORY

We hope in this chapter to explain what it means for a statement (such as the Continuum Hypothesis, CH) to be independent of a theory (such as ZFC), and how independence can be proved by the construction of suitable models. Our theme is that models of set theory are mathematical objects which we can talk about, in the language of naïve set theory, just as we talk about partially ordered sets or Banach algebras: we do not formally need any metamathematical considerations.

This chapter is not a course in mathematical logic, and we shall omit all proofs and discussions which are at all technical.

We first describe the language $\mathscr{L}(\hat{\in},\hat{=})$ of set theory with which we shall work. Our language is, in fact, a "first-order language, with equality", and our introductory development of formal logic will be specific to this language. There are other languages and other, more general, presentations of formal logic: see [23], for example.

Informally, $\mathscr{L}(\hat{\in},\hat{=})$ consists of certain expressions, possibly involving the symbols $\hat{\in}$ and $\hat{=}$ and some "variables" x,y,z,... : '$x \mathbin{\hat{=}} y$', '$\forall z(\exists x(x \mathbin{\hat{=}} z))$', and '$z \mathbin{\hat{\in}} z$' are typical elements of $\mathscr{L}(\hat{\in},\hat{=})$.

Formally, we define $\mathscr{L}(\hat{\in},\hat{=})$ as follows. The <u>alphabet</u> of $\mathscr{L}(\hat{\in},\hat{=})$ is an infinite set of positive integers, the first seven of which are denoted by

$$'\vee', \quad '\neg', \quad '(', \quad ')', \quad '\exists', \quad '\hat{=}', \quad \text{and} \quad '\hat{\in}',$$

respectively. The remaining elements of the alphabet are

called variable symbols, and are denoted in increasing order by $x_1, x_2, x_3, \ldots$ or by $x, y, z, \ldots$. Intuitively, $\vee$ means "or", $\neg$ means "not", $\exists$ means "there exists", $\hat{=}$ denotes equality, and $\hat{\in}$ denotes membership. However, $\hat{=}$ and $\hat{\in}$ have no *necessary* interpretation as equality or membership; it is to stress this that we use the symbols $\hat{=}$ and $\hat{\in}$, rather than $=$ and $\in$, respectively.

The language $\mathscr{L}(\hat{\in},\hat{=})$ consists of certain finite sequences $\langle a_1,\ldots,a_n\rangle$, where $a_1,\ldots,a_n$ belong to the alphabet. These finite sequences are the formulae of the language; the length of $\langle a_1,\ldots,a_n\rangle$ is n. The syntax of the language specifies the sequences which are the formulae: it is a list of rules, to be given below, which state how formulae are formed. In stating these rules, and throughout the book, we shall use some conventions to abbreviate and to make more comprehensible the formulae that we use: for example, we write '$x_1 \hat{\in} x_2$' for $\langle x_1, \hat{\in}, x_2\rangle$.

Here are the rules of the syntax.

1 ATOMIC RULE
For all variable symbols x_i and x_j, '$x_i \hat{=} x_j$' and '$x_i \hat{\in} x_j$' are formulae. These are the atomic formulae.

2 CONNECTIVE RULE
If ϕ and ψ are formulae, then '$(\phi) \vee (\psi)$', and '$\neg(\phi)$' are each formulae.

3 QUANTIFIER RULE
If ϕ is a formula and x_i is a variable symbol, then '$\exists x_i(\phi)$' is a formula.

In Rule 3, we do not require that the variable symbol x_i necessarily occurs in the formula ϕ.

The language $\mathscr{L}(\hat{\in},\hat{=})$ is the smallest set of finite sequences of the alphabet which is closed under applications of any of the above three rules. For example, '$(\hat{=})$' is not a formula of $\mathscr{L}(\hat{\in},\hat{=})$.

We introduce a number of standard abbreviations:

$\phi \wedge \psi$ for $\neg((\neg(\phi)) \vee (\neg(\psi)))$,

$\forall x_i(\phi)$ for $\neg(\exists x_i(\neg(\phi)))$,

$\phi \Rightarrow \psi$ for $((\neg(\phi)) \vee (\psi))$,

$\phi \Leftrightarrow \psi$ for $(\phi \Rightarrow \psi) \wedge (\psi \Rightarrow \phi)$,

$x_i \,\hat{\notin}\, x_j$ for $\neg(x_i \,\hat{\in}\, x_j)$.

Intuitively, $\wedge$ means "and", $\forall$ means "for all", $\Rightarrow$ means "implies", and $\Leftrightarrow$ means "if and only if".

Henceforth, we shall drop some parentheses, and we shall write $\mathscr{L}$ for $\mathscr{L}(\hat{\in}, \hat{=})$ and "variable" for "variable symbol".

Let ϕ be a formula of $\mathscr{L}$. A <u>subformula</u> of ϕ is a formula ψ which, as a sequence, is formed by taking in order some consecutive elements of the sequence ϕ. For example, consider the formula

$$(\exists x_5(x_2 \,\hat{\in}\, x_7)) \vee (\exists x_7(x_1 \,\hat{=}\, x_7)). \qquad (1)$$

Then $\exists x_5(x_2 \,\hat{\in}\, x_7)$ and $x_1 \,\hat{=}\, x_7$ are subformulae of (1), but $\exists x_5(x_1 \,\hat{=}\, x_7)$ is not a subformula. It is fairly easy to see that the subformulae of a formula ϕ are precisely those that arise in the construction of ϕ. Further, each formula is uniquely defined from (either one or two) uniquely specified subformulae of smaller length by a unique rule of formation. We shall frequently use "induction on the length of formulae" to prove a result for all formulae: our remarks on the formation of formulae show that this is a valid procedure.

Let x_i be a variable, and let ϕ be a formula of $\mathscr{L}$. The <u>scope</u> of a particular occurrence of $\exists x_i$ in ϕ is the unique subformula of ϕ which, as a subsequence of ϕ, begins with that occurrence of $\exists x_i$. For example, the scope of the single occurrence of $\exists x_5$ in the formula (1), above, is $\exists x_5(x_2 \,\hat{\in}\, x_7)$. An occurrence of a variable x_i in a formula ϕ is <u>free</u> if it does not belong to the scope of any occurrence of $\exists x_i$ in ϕ. Otherwise, that occurrence of the variable x_i is <u>bound</u>. For example, the first occurrence of x_7 in (1) is free, but the second and third occurrences are bound. A variable x_i is a <u>free variable</u> in ϕ if there is at least one free occurrence of x_i in ϕ, and x_i is a <u>bound variable</u> if every occurrence of x_i in ϕ is bound. Intuitively, a formula expresses a property of its free variables. We write $\phi(x_1,\dots,x_n)$ to indicate a formula all of whose free variables belong to the set $\{x_1,\dots,x_n\}$; of course, not all the variables need actually occur. However, when no variables are displayed after a formula, no commitment concerning the set of free variables is made: for example, a formula ϕ can have x_2 as a free variable, but $\phi(x_1)$ cannot.

A <u>sentence</u> is a formula which has no free variables. Intuitively, a sentence is an assertion which is either true or false. A <u>theory</u> in $\mathscr{L}$ is a (possibly empty, possibly infinite) set of sentences of $\mathscr{L}$. Let S and T be two theories in $\mathscr{L}$, and let ϕ be a sentence of $\mathscr{L}$. We write

$$S + T \text{ for } S \cup T, \text{ and } T + \phi \text{ for } T + \{\phi\}.$$

We now explain how a theory "proves" a formula.

The language $\mathscr{L}$ contains certain formulae which we deem to be the <u>logical axioms</u> of $\mathscr{L}$. There is some freedom in the choice of these axioms, and we do not wish to explicitly write down our choice: let us adopt, for example, the list given in [23, page 104]. The logical axioms are all "trivially true". They include '$\phi \Rightarrow \phi$', '$\phi \Rightarrow (\psi \Rightarrow (\phi \wedge \psi))$', and '$\forall x_i(\phi) \Rightarrow \phi$'; perhaps more controversially, '$\phi \vee (\neg\phi)$', the law of the excluded middle, is a logical axiom. The set of logical axioms also includes those formulae which show that $\hat{=}$ is an equivalence relation which $\hat{\in}$ respects: for example, '$x \mathrel{\hat{=}} x$' and '$(x \mathrel{\hat{=}} y) \wedge (y \mathrel{\hat{\in}} z) \Rightarrow (x \mathrel{\hat{\in}} z)$' are logical axioms.

A <u>rule of inference</u> is a procedure for deriving

another formula from a collection of formulae of the language. In fact, we have only one rule of inference, modus ponens, in our formulation.

Modus ponens: from $\{\phi, \phi \Rightarrow \psi\}$, derive ψ.

Some accounts of logic contain a second rule of inference, that of generalization: from $\{\phi\}$, derive $\forall x \phi$. This rule is unnecessary if the logical axioms are suitably chosen, as they are in [23].

4.1 DEFINITION

Let T be a theory in $\mathscr{L}$, and let ϕ be a formula of $\mathscr{L}$. Then T proves ϕ (in $\mathscr{L}$) if there is a finite sequence $\langle\phi_1,\ldots,\phi_n\rangle$ of formulae of $\mathscr{L}$ such that $\phi_n = \phi$ and such that each ϕ_i is either an element of T, or a logical axiom, or is obtained from two formulae of $\{\phi_1,\ldots,\phi_{i-1}\}$ by modus ponens. In this case, $\langle\phi_1,\ldots,\phi_n\rangle$ is a proof of ϕ from T.

We write

$$T \vdash \phi \text{ (respectively, } T \nvdash \phi)$$

if T proves ϕ (respectively, if it is not the case that T proves ϕ).

We emphasize that the notion of "proof" in Definition 4.1 is an abstract mathematical relation which does not necessarily correspond to any intuitive notion of proof. A formula either is or is not proved by a theory; an element of an algebra either is or is not invertible. We do not imply, when making the definition, that there is any algorithm that determines whether $T \vdash \phi$ or $T \nvdash \phi$.

4.2 DEFINITION

A formula ϕ is logically derivable if $\emptyset \vdash \phi$, where $\emptyset$ is the empty set of sequences: we write

$\vdash \phi$.

Two formulae ϕ and ψ are <u>equivalent</u> in a theory T if $T \vdash (\phi \Leftrightarrow \psi)$, and ϕ and ψ are <u>logically equivalent</u> if $\vdash (\phi \Leftrightarrow \psi)$.

Let ϕ be a formula. A sentence formed from ϕ by successively inserting at the beginning $\forall x_i$ for each free variable x_i which occurs in ϕ is a <u>universal closure</u> of ϕ : for example, the sentences $\forall x_1 \forall x_2 (x_1 \hat{=} x_2)$ and $\forall x_2 \forall x_1 (x_1 \hat{=} x_2)$ are both universal closures of $x_1 \hat{=} x_2$. It is routine, given the logical axioms, to show that any two universal closures of ϕ are logically equivalent, and so we may refer to ϕ^*, <u>the</u> universal closure of ϕ.

If a formula ϕ is logically derivable, then its universal closure is also logically derivable. More generally, if T is a theory and if $T \vdash \phi$, then $T \vdash \phi^*$, and we can arrange that the proof consists only of sentences: to show this is again a routine exercise ([23, §2.4]).

Suppose that $\langle\phi_1, \ldots, \phi_m\rangle$ and $\langle\psi_1, \ldots, \psi_n\rangle$ are proofs from a theory T. Then $\langle\phi_1, \ldots, \phi_m, \psi_1, \ldots, \psi_n\rangle$ is a proof from T. It is now clear that, if $T \vdash \phi$ and $T \vdash \psi$, then $T \vdash \phi \wedge \psi$: use the logical axiom '$\phi \Rightarrow (\psi \Rightarrow (\phi \wedge \psi))$' and modus ponens twice. It is also true that $T + \phi \vdash \psi$ if and only if $T \vdash (\phi \Rightarrow \psi)$.

4.3 DEFINITION

Let T be a theory in $\mathscr{L}$. Then T is <u>inconsistent</u> if there is a sentence ϕ such that $T \vdash \phi \wedge (\neg\phi)$; otherwise, T is <u>consistent</u>.

If T is inconsistent, then $T \vdash \phi$ for each sentence ϕ, and so T has little interest; we prefer consistent theories. Note that $T \vdash \phi$ if and only if $T + (\neg\phi)$ is inconsistent. Equivalently, $T \nvdash \phi$ if and only if $T + (\neg\phi)$ is consistent.

We write

Con T

if the theory T is consistent. A sentence ϕ is <u>relatively consistent</u> with a theory T if

Con T implies Con(T + ϕ).

i.e., either T is inconsistent or T + ϕ is consistent. (Note that every sentence is relatively consistent with an inconsistent theory.) We shall remark shortly that it cannot be proved that ZFC is consistent, and so it is natural to consider whether or not a given sentence ϕ is relatively consistent with ZFC.

We now come to the notions of model and satisfaction, which will be central to much of our later work.

4.4 DEFINITION

A <u>model</u> $\mathfrak{M}$ is a pair (M,E), where M is a non-empty set and E is a subset of $M \times M$. The set M is the <u>underlying set</u> of $\mathfrak{M}$. The model $\mathfrak{M}$ is <u>countable</u> if M is countable.

Throughout, we shall adopt the convention that a model is denoted by an upper-case fraktur letter and that the underlying set is denoted by the same roman letter.

The subset E of $M \times M$ can be regarded as a binary relation on M : aEb if and only if $(a,b) \in E$.

Formally, our definition of a model has no connection with anything preceding the definition. However, it is formulated with a connection to the language $\mathscr{L}$ in mind (and more general definitions of model would be appropriate if we were considering more general languages). Our "model" is sometimes termed an "$\mathscr{L}$-structure" in the literature.

4.5 DEFINITION

Let $\mathfrak{M} = (M, E_M)$ and $\mathfrak{N} = (N, E_N)$ be models. Then $\mathfrak{M}$ is a <u>submodel</u> of $\mathfrak{N}$, and $\mathfrak{N}$ is an <u>extension</u> of $\mathfrak{M}$, if $M \subset N$ and if $E_N \cap (M \times M) = E_M$.

For example, $(\mathbb{R}, <)$ is a model, and $(\mathbb{Q}, <)$ is a submodel of $(\mathbb{R}, <)$.

Let $\mathfrak{M} = (M, E)$ be a model, and let $\mathscr{L} = \mathscr{L}(\hat{\epsilon}, \hat{=})$ be the language of set theory that we are employing. Consider a formula $\phi = \phi(x_1, \ldots, x_n)$ of $\mathscr{L}$ and elements $a_1, \ldots, a_n$ of M. Then there is a natural interpretation of ϕ as a statement (relative to $\mathfrak{M}$) about the elements $a_1, \ldots, a_n$. Informally, we interpret $\hat{\epsilon}$ as E, $\hat{=}$ as equality in M, and x_i as a_i. More formally, we define by induction on the length of formulae the <u>truth of ϕ at</u> $(a_1, \ldots, a_n)$ <u>in</u> $\mathfrak{M}$ by rules to be set out below. We write

$$\mathfrak{M} \models \phi[a_1, \ldots, a_n],$$

and say that $\mathfrak{M}$ <u>satisfies</u> $\phi[a_1, \ldots, a_n]$, if ϕ is true at $(a_1, \ldots, a_n)$ in $\mathfrak{M}$. If ϕ is not true at $(a_1, \ldots, a_n)$ in $\mathfrak{M}$, then we write $\mathfrak{M} \not\models \phi[a_1, \ldots, a_n]$.

The rules should be compared with the rules of syntax in $\mathscr{L}$ given on page 55. The definitions assume that all the variables, bound or free, occurring in ϕ belong to the set $\{x_1, \ldots, x_n\}$.

1 Let x_i and x_j be variables. Then:

$$\mathfrak{M} \models (x_i \mathrel{\hat{=}} x_j)[a_1, \ldots, a_n] \text{ if } a_i = a_j;$$

$$\mathfrak{M} \models (x_i \mathrel{\hat{\epsilon}} x_j)[a_1, \ldots, a_n] \text{ if } a_i E a_j.$$

2 Let ϕ and ψ be formulae. Then:

$$\mathfrak{M} \models (\phi \vee \psi)[a_1, \ldots, a_n] \text{ if } \mathfrak{M} \models \phi[a_1, \ldots, a_n]$$
$$\text{or if } \mathfrak{M} \models \psi[a_1, \ldots, a_n];$$

$\mathfrak{M} \models (\neg\phi)[a_1,\ldots,a_n]$ if $\mathfrak{M} \not\models \phi[a_1,\ldots,a_n]$.

3 Let ϕ be a formula, and let x_i be a variable. Then:

$\mathfrak{M} \models (\exists x_i\phi)[a_1,\ldots,a_n]$ if, for some $b \in M$,

$$\mathfrak{M} \models \phi[a_1,\ldots,a_{i-1}, b, a_{i+1},\ldots,a_n].$$

It follows from the definitions that, if ϕ is a formula and x_i is a variable, then:

$\mathfrak{M} \models (\forall x_i\phi)[a_1,\ldots,a_n]$ if, for all $b \in M$,

$$\mathfrak{M} \models \phi[a_1,\ldots,a_{i-1}, b, a_{i+1},\ldots,a_n].$$

If x_i is a bound variable of ϕ, then the truth of ϕ at $(a_1,\ldots,a_n)$ does not depend on a_i, and so we can suppose that $\mathfrak{M} \models \phi[a_1,\ldots,a_n]$ is defined whenever the free variables of ϕ belong to $\{x_1,\ldots,x_n\}$. In particular, if ϕ is a sentence, then the truth of ϕ in $\mathfrak{M}$ is well defined: if ϕ is true, then we write $\mathfrak{M} \models \phi$, and say that $\mathfrak{M}$ is a model of ϕ.

Notice that $\mathfrak{M} \models \phi$ if and only if $\mathfrak{M} \not\models \neg\phi$.

4.6 DEFINITION

Let $\mathfrak{M}$ be a model. The theory of $\mathfrak{M}$ is the set of sentences ϕ in $\mathscr{L}$ such that $\mathfrak{M} \models \phi$, and this set is denoted by $Th(\mathfrak{M})$; $\mathfrak{M}$ is a model of a theory T if $T \subset Th(\mathfrak{M})$, and we then write $\mathfrak{M} \models T$.

4.7 DEFINITION

Let $\mathfrak{M} = (M,E_M)$ and $\mathfrak{N} = (N,E_N)$ be models. Then $\eta : \mathfrak{M} \to \mathfrak{N}$ is an isomorphism if η is a bijection from M onto N such that aE_Mb if and only if $\eta(a)E_N\eta(b)$. In this case $\mathfrak{M}$ and $\mathfrak{N}$ are isomorphic, and we write $\mathfrak{M} \cong \mathfrak{N}$.

Let $\eta : \mathfrak{M} \to \mathfrak{N}$ be an isomorphism, let $\phi(x_1,\ldots,x_n)$

be a formula, and let $a_1,\dots,a_n \in M$. Then
$\mathfrak{M} \models \phi[a_1,\dots,a_n]$ if and only if $\mathfrak{N} \models \phi[\eta(a_1),\dots,\eta(a_n)]$.
Hence $Th(\mathfrak{M}) = Th(\mathfrak{N})$.

The sentences in the set of the (previously unspecified) logical axioms of $\mathscr{L}$ are true in every model (see [23, 25A]). Further, modus ponens preserves truth in a model, in the sense that, if ϕ and ψ are sentences of $\mathscr{L}$, if $\mathfrak{M} \models \phi$, and if $\mathfrak{M} \models \phi \rightarrow \psi$, then $\mathfrak{M} \models \psi$. Let T be a theory, and let ϕ be a sentence. It follows that, if $\mathfrak{M} \models T$ and if $T \vdash \phi$, then $\mathfrak{M} \models \phi$. Thus, if T has a model, then T is consistent, for it cannot be that $\mathfrak{M} \models \phi \wedge (\neg\phi)$. The converse of this statement is also true: it is Gödel's completeness theorem, which we state for our language $\mathscr{L}$.

4.8 THEOREM

Let T be a theory of $\mathscr{L}$. Then the following are equivalent:

(a) T is consistent;

(b) T has a model;

(c) T has a countable model. ∎

In symbols, Con T if and only if there is a model $\mathfrak{M}$ such that $\mathfrak{M} \models T$.

That a consistent theory has a countable model is a consequence of the fact that $\mathscr{L}$, and hence T, is countable. The proof of Gödel's theorem which is given in [23, §2.5], for example, provides a countable model for each consistent theory in $\mathscr{L}$. There are some developments of forcing in which the existence of countable models is central, but we shall only use it in some peripheral remarks.

It follows from 4.8 that a sentence ϕ is logically derivable if and only if it is true in every model. More generally, if T is a theory, then $T \vdash \phi$ if and only if each model of T is also a model of ϕ. This is the importance of Gödel's completeness theorem : *it identifies the notions of proof and of truth.*

Our approach to the proof of the relative consistency of a sentence ϕ with ZFC will be to construct from a model of ZFC a model of ZFC + ϕ. It is because we are willing to apply Gödel's completeness theorem that we can step from this to the assertion that ϕ is relatively consistent with ZFC.

Models of quite a number of theories are studied, and we shall give a few rather elementary examples. But the examples that we are seriously interested in here are "models of set theory itself". More precisely, we shall eventually be concerned with models of the theory ZFC : this is Zermelo-Fraenkel set theory with the Axiom of Choice. (There are other set theories that could equally well be considered, such as GB, Gödel-Bernays set theory, but ZFC seems to be the most popular.)

We first list axioms for ZFC. We give a formal version of each axiom, and an informal interpretation: it is not expected that the reader will always easily verify, and it is not always true, that the formal version is a precise formalization of the interpretation. There are a number of variants on this list that are current; we do not wish to discuss the possible variations. Our choice is made for the sake of later technical convenience. The point is that the axioms are an attempt to capture basic intuitions about sets, and a judgement on the success of this attempt must be subjective.

AXIOM 0 Set existence.

$$\exists x(x \mathrel{\hat{=}} x)$$

'There is a set'.

AXIOM 1 Extensionality.

$$\forall x \forall y((x \mathrel{\hat{=}} y) \leftrightarrow \forall z((z \mathrel{\hat{\in}} x) \leftrightarrow (z \mathrel{\hat{\in}} y)))$$

'Two sets are equal if and only if they have the same elements'.

AXIOM 2 Comprehension.

Let $\phi(x_1,\ldots,x_n)$ be a formula in $\mathscr{L}$.

$$\forall x_2 \ldots \forall x_n \forall x_{n+1} \exists x_{n+2} \forall x_1 ((x_1 \mathbin{\hat{\in}} x_{n+2}) \Leftrightarrow ((x_1 \mathbin{\hat{\in}} x_{n+1}) \wedge \phi)$$

'If A is a set and P is a "property", then $\{B \in A : P(B)$ is true$\}$ is a set.'

In fact, Axiom 2 is an infinite scheme of axioms, one axiom corresponding to each formula $\phi(x_1,\ldots,x_n)$. The language $\mathscr{L}$ and the formally stated axiom make precise our intuitive notion of "property".

It follows from Comprehension that there is a set with no elements, and, by Extensionality, that this set is unique. It is denoted by $\emptyset$.

AXIOM 3 Pairing.

$$\forall x \forall y \exists z((x \mathbin{\hat{\in}} z) \wedge (y \mathbin{\hat{\in}} z))$$

'If A and B are sets, then there is a set C with $A \in C$ and $B \in C$.'

It follows from Pairing and Comprehension that, if A and B are sets, then so is $\{A,B\}$.

We introduce the abbreviation $z \subset y$ for

$$\forall t(t \mathbin{\hat{\in}} z \Rightarrow t \mathbin{\hat{\in}} y).$$

AXIOM 4 Union.

$$\forall x \exists y \forall z(z \mathbin{\hat{\in}} x \Rightarrow z \subset y)$$

'If A is a set, then there is a set containing the members of the members of A.'

It follows from Union and Comprehension that, if A is a set, then so is $\bigcup\{B : B \in A\}$; we often denote this latter set by $\bigcup A$.

AXIOM 5 Power set.

$\forall x \exists y \forall z (z \subset x \Rightarrow z \hat{\in} y)$

'If A is a set, then there is a set whose elements include all the subsets of A.'

AXIOM 6 Replacement.

Let $\phi(x_1, x_2)$ be a formula in $\mathscr{L}$.

$\forall x_3 \exists x_4 \forall x_1 ((x_1 \hat{\in} x_3 \wedge \exists x_2(\phi)) \Rightarrow \exists x_2(x_2 \hat{\in} x_4 \wedge \phi))$

The essence of Replacement is the following. (A function F is defined by a property P if $F(A) = B$ whenever $P(A,B)$ holds.)

'The image of a set under a function defined by a property is itself a set.'

Our formulation of Replacement is a slight variant of this, but each formulation can be proved from the other by using Axioms 0 - 5, and the Axiom of Regularity given below. Axiom 6 is also an infinite scheme of axioms.

AXIOM 7 Regularity (or Foundation).

$\forall x \exists y \forall z ((z \hat{\notin} x) \vee (y \hat{\in} x \wedge (z \hat{\in} y \Rightarrow z \hat{\notin} x)))$

'Each non-empty set A contains an element B such that there are no elements belonging to both B and A.'

It follows from Regularity and Comprehension that

$A \in A$ is prohibited. This fact is used in the theory of ordinals, to be given in Chapter 5.

AXIOM 8 <u>Infinity</u>

$$\exists x \forall y \exists z((z \hat{\in} x) \wedge (y \hat{\in} x \rightarrow y \hat{\in} z))$$

'There is a non-empty set A such that each element of A is an element of an element of A.'

This axiom is often stated in a slightly different form: 'There is a set W such that the empty set belongs to W and such that $A \cup \{A\}$ belongs to W whenever A is an element of W.' This form can be deduced from our version by using the other axioms. The purpose of the axiom is to establish the existence of an infinite set: the set W contains $\emptyset$, $\{\emptyset\}$, $\{\emptyset,\{\emptyset\}\}$, $\{\emptyset,\{\emptyset\},\{\emptyset,\{\emptyset\}\}\},\ldots$, and hence W "really" contains $\mathbb{Z}^+$.

AXIOM 9 <u>Axiom of Choice AC</u>

$$\forall x \exists y \exists z(x \hat{=} \emptyset \vee (z \hat{\in} x \wedge z \hat{=} \emptyset) \vee$$
$$(y \hat{\in} x \wedge z \hat{\in} x \wedge y \hat{\neq} z \wedge y \cap z \hat{\neq} \emptyset) \vee$$
$$(\forall t \exists u \forall v(t \hat{\notin} x \vee (u \hat{\in} t \wedge u \hat{\in} y \wedge (v \hat{\notin} t \vee v \hat{\notin} y \vee v \hat{=} u))))),$$

where '$x \hat{=} \emptyset$' abbreviates '$\forall y(y \hat{\notin} x)$' and '$y \cap z \hat{=} \emptyset$' abbreviates '$\forall x(x \hat{\notin} y \wedge x \hat{\notin} z)$'.

'For each non-empty set A of pairwise disjoint, non-empty sets, there is a set B such that the intersection of B with each member of A contains exactly one element.'

Assuming Axioms 1 - 8, AC is equivalent to 'For each non-empty set A, there is a function $f : A \rightarrow \bigcup A$ such that $f(B) \in B$ for each $B \in A$ with $B \neq \emptyset$.' Such a function f is a <u>choice function</u> for A.

It is manifest that the formal version of AC is cumbersome and non-enlightening.

4.9 DEFINITION

The theory consisting of the set of sentences which are proved by Axioms 0 - 8 is ZF, and the theory consisting of the set of sentences which are proved by Axioms 0 - 9 is ZFC.

Thus ZFC is an infinite theory in the language $\mathscr{L}$.

We write "$\mathfrak{M} \models$ Comprehension" if a model $\mathfrak{M}$ satisfies each sentence of the infinite axiom scheme of Comprehension, etc.

Our point of view is that models (and, in particular, models of ZFC) are just algebraic objects. They have the same ontological status as partially ordered sets, for example. Necessarily a model of the whole of ZFC will be complicated, but there are familiar algebraic objects which are models of some subtheories of the theory ZFC: we give some examples.

4.10 EXAMPLES

(i) Let $\mathfrak{M} = (\mathbb{R}, <)$. Then

$$\mathfrak{M} \models \text{Set existence + Extensionality + Pairing}$$
$$\text{+ Union + Infinity + Power set + Replacement.}$$

For example, $\mathfrak{M} \models$ Pairing because, given $a, b \in \mathbb{R}$, there exists $c \in \mathbb{R}$ such that $a < c$ and $b < c$. Note that $b \subset a$ if and only if $b \leqslant a$. Thus $\mathfrak{M} \models$ Power set because, given $a \in \mathbb{R}$, there exists $b \in \mathbb{R}$ such that $c < b$ whenever $c \leqslant a$ in $\mathbb{R}$. The model $\mathfrak{M}$ does not satisfy the other axioms. For example, $\mathfrak{M} \not\models$ Regularity, for, given $a \in \mathbb{R}$, it is not true that there exists $b < a$

such that, for each $c \in \mathbb{R}$, if $c < b$, then $c \not< a$.

So already we have a model that satisfies seven out of ten axioms.

(ii) Let $\mathfrak{M} = (\mathbb{N}, <)$. Then

$$\mathfrak{M} \models \text{Regularity}.$$

Let $a \in \mathbb{N}$. If $a = 1$, then $c \not< a$ for each $c \in \mathbb{N}$ ("a is empty"). If $a \geq 2$, then $1 < a$ ("a is non-empty, and $b = 1$ is the element such that no element of b is an element of a").

(iii) Let $\mathfrak{M} = (\mathbb{Z}^+, E)$, where, for $m,n \in \mathbb{Z}^+$, mEn if the mth digit in the binary expansion of n is 1, i.e., if $n = \sum_{i=o}^{\infty} n_i 2^i$, where $n_i \in \{0,1\}$, then mEn if $n_m = 1$. Then $\mathfrak{M}$ satisfies each of the above axioms except for Infinity, but it is not entirely straightforward to check this: it is easier to do this if one realizes that $\mathfrak{M}$ is isomorphic to $(V_\omega, \in)$, to be defined in Chapter 5.

So we obtain a model which satisfies nine out of ten axioms.

The theory generated by all the axioms save for Infinity can be regarded as a reformulation of number theory; the Axiom of Infinity takes us from number theory to set theory.

4.11 DEFINITION

Let $\mathfrak{M} = (M,E)$ be a model, let $k \in \mathbb{N}$, and let S be a subset of M^k. Then S is a <u>simple</u> subset if there is a formula $\phi(x_1,\ldots,x_k,x_{k+1},\ldots,x_{k+n})$ and elements $c_1,\ldots,c_n$ in M such that

$$S = \{(b_1,\ldots,b_k) \in M^k : \mathfrak{M} \models \phi[b_1,\ldots,b_k,c_1,\ldots,c_n]\}.$$

Intuitively, the simple subsets are those sets which can be defined by a formula in the language $\mathscr{L}$; the formula is allowed to have parameters. Indeed, the simple sets are often called the <u>definable</u> sets.

For example, let us determine the simple subsets of $\mathbb{R}$ in the model $(\mathbb{R},<)$. Let $S \subset \mathbb{R}$ be simple. Then there is a formula $\phi(x_1,\ldots,x_{n+1})$ and $t_1,\ldots,t_n \in \mathbb{R}$ such that

$$S = \{b \in \mathbb{R} : (\mathbb{R},<) \models \phi[b,t_1,\ldots,t_n]\}.$$

We can suppose that $t_1 < t_2 < \ldots < t_n$. Let Π be the set of order-isomorphisms $\pi : \mathbb{R} \to \mathbb{R}$ such that $\pi(t_j) = t_j$ $(j = 1,\ldots,n)$. Set $b \sim c$ in S if there exists $\pi \in \Pi$ with $\pi(b) = c$. Then $\sim$ is an equivalence relation on $\mathbb{R}$, and the equivalence classes are the intervals $(-\infty,t_1)$, $\{t_1\}$, (t_1,t_2), ... For each $\pi \in \Pi$, $b \in S$ if and only if $\pi(b) \in S$, and so $\pi(S) = S$ $(\pi \in \Pi)$. Thus S is a union of equivalence classes, and so each simple subset of $\mathbb{R}$ is a finite union of intervals.

Our definition of "simple sets" involves the notion of satisfaction. We now give an algebraic characterization of these sets.

Let $\mathfrak{M} = (M,E)$ be a model, and let $k \in \mathbb{N}$. The <u>basic</u> sets of M^k are

$$A^k_{ij} = \{(a_1,\ldots,a_k) \in M^k : a_i E a_j\},$$

and

$$B^k_{ij} = \{(a_1,\ldots,a_k) \in M^k : a_i = a_j\},$$

where $i,j \in \{1,\ldots,k\}$. Define

$$\pi_k : (a_1,\ldots,a_{k+1}) \mapsto (a_1,\ldots,a_k), \quad M^{k+1} \to M^k.$$

Consider the collection of families F such that $F = \bigcup\{F_k : k \in \mathbb{N}\}$, and the subfamilies $F_k \subset P(M^k)$ satisfy the following conditions for each $k \in \mathbb{N}$:

(i) $A_{ij}^k, B_{ij}^k \in F_k \quad (i,j = 1,\ldots,k)$;

(ii) $A \cup B \in F_k \quad (A,B \in F_k)$;

(iii) $M^k \setminus A \in F_k \quad (A \in F_k)$;

(iv) $\pi_k(A) \subset F_k \quad (A \in F_{k+1})$.

There is a family with these properties, and the intersection of all the families with these properties has the properties : this intersection is the family of simple sets. The proof of this claim is by induction on the length of the formulae involved : of course, union corresponds to $\vee$, complementation to $\neg$, and projection to $\exists$.

Let $\mathfrak{M}$ be a model. We now claim that the statement '$\mathfrak{M} \models$ ZFC' can be given an algebraic interpretation which has no reference to the language $\mathcal{L}$. To do this, we should interpret '$\mathfrak{M} \models \phi$' for each of the above-specified axioms of ZFC. In fact, we shall do this for four of the axioms; this should be sufficient to convince the reader that it can be done for all of the axioms. We first give a definition that will also be required in Chapter 7.

4.12 DEFINITION

Let $\mathfrak{M} = (M,E)$ be a model. For $a \in M$, set

$$\tilde{a} = \{b \in M : bEa\}.$$

Clearly each $\tilde{a}$ is a simple subset of M.

4.13 EXAMPLE

Let $\mathfrak{M}$ be a model.

(i) $\mathfrak{M} \models$ Extensionality.

This is equivalent to the condition that, if $a,b \in M$ and $\tilde{a} = \tilde{b}$, then $a = b$. In this case, the map $a \mapsto \tilde{a}$ is an injection.

(ii) $\mathfrak{M} \models$ Comprehension.

This is equivalent to the condition that, if $a \in M$ and if S is a simple subset of M, then there exists $b \in M$ such that $\tilde{b} = S \cap \tilde{a}$.

(iii) $\mathfrak{M} \models$ Infinity.

An element b of a set $\tilde{a}$ is an <u>E-maximal</u> element if, for each $c \in \tilde{a}$, it is not true that bEc. The above statement is equivalent to the condition that, for some $a \in M$, $\tilde{a}$ is non-empty and has no E-maximal element. Note that it is not correct merely to require that $\tilde{a}$ be infinite for some $a \in M$. For example, observe that

$$(\mathbb{N}^{\mathbb{N}}, <_F) \not\models \text{Infinity},$$

although $\tilde{a}$ is infinite for most $a \in \mathbb{N}^{\mathbb{N}}$.

(iv) $\mathfrak{M} \models$ Regularity.

An element b of a set $\tilde{a}$ is an <u>E-minimal</u> element if, for each $c \in \tilde{a}$, it is not true that cEb. The above statement is equivalent to the condition that, for each $a \in M$, either $\tilde{a} = \emptyset$ or $\tilde{a}$ has an E-minimal element. Note that this condition does not imply that each non-empty subset of M, has an E-minimal element. For example, $(\mathbb{N}^{\mathbb{N}}, <_F) \models$ Regularity, but the subset $\{f \in \mathbb{N}^{\mathbb{N}} : f(n) \to \infty$ as $n \to \infty\}$ does not have an E-minimal element.

We have indicated that we are only interested in considering consistent theories. Another, apparently desirable, feature of a theory is that it be complete, in the sense of the following definition.

Let T be a theory in the language $\mathscr{L}$. We write

$$\text{Thm}(T) = \{\phi \in \mathscr{L} : \phi \text{ is a sentence and } T \vdash \phi\}.$$

4.14 DEFINITION

Let T be a theory in $\mathscr{L}$. Then T is <u>complete</u> if, for each sentence ϕ, either $\phi \in \text{Thm}(T)$ or

$\neg\phi \in \mathrm{Thm}(T)$. A sentence ϕ is <u>independent of T</u> if $\phi \notin \mathrm{Thm}(T)$ and $\neg\phi \notin \mathrm{Thm}(T)$.

Thus ϕ is independent of T if T neither proves ϕ nor $\neg\phi$, and, if T is consistent, then ϕ is independent of T if and only if both ϕ and $\neg\phi$ are relatively consistent with T.

Let $\mathfrak{M}$ be a model. Then $\mathrm{Th}(\mathfrak{M})$ is a complete, consistent theory. Gödel's completeness theorem 4.8 shows that, if T is a consistent theory, then

$$\mathrm{Thm}(T) = \bigcap\{\mathrm{Th}(\mathfrak{M}) : \mathfrak{M} \text{ is a model and } \mathfrak{M} \models T\}.$$

We give an example of a complete, consistent theory.

4.15 EXAMPLE

Let T be the set of axioms for a dense total order without endpoints. Thus T consists of:

(0) $\exists x(x \mathrel{\hat{=}} x)$;

(1) $\forall x\forall y\forall z((x \mathrel{\hat{\in}} y \wedge y \mathrel{\hat{\in}} z) \rightarrow (x \mathrel{\hat{\in}} z))$;

(2) $\forall x(x \mathrel{\hat{\notin}} x)$;

(3) $\forall x\forall y(x \mathrel{\hat{\in}} y \vee x \mathrel{\hat{=}} y \vee y \mathrel{\hat{\in}} x)$;

(4) $\forall x\forall y\exists z(x \mathrel{\hat{\in}} y \rightarrow (x \mathrel{\hat{\in}} z \wedge z \mathrel{\hat{\in}} y))$.

(5) $\forall x\exists y(x \mathrel{\hat{\in}} y)$;

(6) $\forall x\exists y(y \mathrel{\hat{\in}} x)$.

The theory T is consistent because T has a model: $(\mathbb{Q},<) \models T$, and so $(\mathbb{Q},<) \models \mathrm{Thm}(T)$.

Suppose, if possible, that, T is not complete. Then there is a sentence ϕ such that both $T + \phi$ and $T + \neg\phi$ are consistent. By Theorem 4.8, there are countable models $\mathfrak{M}$ and $\mathfrak{N}$ such that $\mathfrak{M} \models T + \phi$ and $\mathfrak{N} \models T + \neg\phi$.

But each countable model of T is just a countable, dense total order without endpoints, and hence is order-isomorphic to $(\mathbb{Q},<)$. Thus $\mathfrak{M}$ and $\mathfrak{N}$ are isomorphic models, a contradiction because $\mathfrak{M} \models \phi$ and $\mathfrak{N} \models \neg\phi$. Hence T is complete.

The distinction between a consistent theory and a complete, consistent theory is related to one that we have already encountered, that between filters and ultrafilters on a Boolean algebra. The details of how this is done are not important for us, and so we shall only give a sketch of the correspondence.

Let S be the set of sentences of the language $\mathscr{L}$. For $\phi, \psi \in S$, set $\phi \sim \psi$ if ϕ and ψ are logically equivalent. Then $\sim$ is an equivalence relation on S. The operations given by $\wedge$, $\vee$, and $\neg$ respect this equivalence relation, and so $\mathfrak{B} = (S/\sim, \wedge, \vee, \neg, [\phi \wedge (\neg\phi)], [\phi \vee (\neg\phi)])$ is a Boolean algebra. It is called the <u>Lindenbaum algebra</u>. The associated order is given by:

$$[\phi] \leqslant [\psi] \quad \text{if} \quad \vdash \phi \Rightarrow \psi \ .$$

Let T be a consistent theory. If $\phi \sim \psi$ in S, then $\phi \in \text{Thm}(T)$ if and only if $\psi \in \text{Thm}(T)$. If $\phi \in \text{Thm}(T)$ and if $\phi \leqslant \psi$, then $\psi \in \text{Thm}(T)$. If $\phi, \psi \in \text{Thm}(T)$, then $\phi \wedge \psi \in \text{Thm}(T)$. Thus Thm(T) is a proper filter in the Boolean algebra $\mathfrak{B}$. The condition that the theory T be complete is clearly the same as the condition that Thm(T) be an ultrafilter in $\mathfrak{B}$.

There are clearly two fundamental questions about the theory ZFC:

(1) *Is* ZFC *consistent?*

(2) *Is* ZFC *complete?*

If ZFC were complete, then mathematics as we know it would reduce to a "finite search". For suppose that ϕ is a sentence. To decide whether or not ϕ were a theorem of ZFC, one would simply search through all proofs from ZFC - they form a countable set - until a proof of either ϕ or of $\neg\phi$ were found.

Fortunately, life is not this boring. The theory ZFC is not complete unless ZFC is inconsistent. Nor can we achieve completeness by adding to ZFC a list of well-chosen additional axioms: this is the content of Gödel's first incompleteness theorem. Let T_o be the set of sentences proved by Axioms 0 - 7 and 9 (cf. Example 4.10(iii)).

4.16 THEOREM

Let T be a theory containing T_o that can be enumerated by an algorithm. Assume that T is consistent. Then there is a sentence of $\mathscr{L}$ which is independent of T. ∎

The possibility that ZFC is consistent is dealt with by Gödel's second incompleteness theorem. For this, it must be accepted that, if T is a theory containing T_o that can be enumerated by an algorithm, then there is a sentence ϕ_T in $\mathscr{L}$ which expresses Con T.

4.17 THEOREM

Let T be a theory containing T_o that can be enumerated by an algorithm. Assume that T is consistent. Then $T \nvdash \phi_T$. ∎

The clause "can be enumerated by an algorithm" is merely our means of avoiding an explicit statement of the conditions that T must satisfy: the theories ZF and ZFC satisfy this clause. Thus, by 4.17, Con ZFC can only be proved in ZFC if ZFC is inconsistent. Nevertheless, we subscribe to the doctrine that ZFC is consistent.

It was a programme of Hilbert of 1920 to secure the foundations of mathematics, essentially by proving that

ZFC is complete and consistent. The theorems of Gödel, announced in 1930, show that this programme must fail.

The proof of Gödel's first incompleteness theorem manufactures - by means of Gödel numbering, see below - a statement which is independent of T. However, the statements which are obtained by this method are neither exciting nor natural. One might still hope that all "naturally arising" sentences are not independent of ZFC, in which case the failure of Hilbert's programme would not seem to be very important. We shall be concerned with the independence from ZFC of two sentences: firstly, with the Continuum Hypothesis CH, and, secondly, with a sentence called NDH ("no discontinuous homomorphisms") that arose from the considerations given in Chapter 1.

4.18 DEFINITION

The sentence NDH is:

For each compact space X, *each homomorphism from* $C(X,\mathbb{C})$ *into a Banach algebra is continuous.*

We hope that the reader will find at least one of CH and NDH to be both exciting and natural.

We shall eventually prove that CH, $\neg$CH, and NDH are all relatively consistent with ZFC. Thus, if ZFC is consistent, then CH is independent of ZFC. Further, modulo Theorem 1.10, we shall prove that NDH is independent of ZFC.

Let ϕ be a sentence of $\mathscr{L}$. We discuss two possible approaches to the problem of proving that ϕ is independent of ZFC.

The first approach is <u>number-theoretic</u>. In essence, the question of the independence of ϕ from ZFC is a question in number theory. It was precisely this insight that Gödel exploited in the proof of his incompleteness theorems.

Define a map $\pi : \mathscr{L}(\hat{\in},\hat{=}) \to \mathbb{N}$ as follows. First assign numbers to the alphabet of $\mathscr{L}(\hat{\in},\hat{=})$:

$$\text{'}\vee\text{'} \mapsto 1,\ \text{'}\neg\text{'} \mapsto 2,\ \text{'('} \mapsto 3,\ \text{')'} \mapsto 4,$$
$$\text{'}\exists\text{'} \mapsto 5,\ \text{'}\hat{=}\text{'} \mapsto 6,\ \text{'}\hat{\in}\text{'} \mapsto 7, \text{ and}$$
$$\text{'}x_n\text{'} \mapsto 7 + n,$$

where $x_1, x_2, \ldots,$ is the enumeration of the variable symbols of $\mathscr{L}$. Using this assignation, we associate to each formula ϕ a finite sequence, say $\langle n_1, \ldots, n_k \rangle$, of natural numbers. Define $\pi(\phi) = p_1^{n_1} \ldots p_k^{n_k}$, where p_i is the ith prime number, and write $\ulcorner\phi\urcorner$ for $\pi(\phi)$; $\ulcorner\phi\urcorner$ is the <u>Gödel number</u> of the formula ϕ. For example,

$$\ulcorner \exists x_1(\neg(x_1 \hat{=} x_1)) \urcorner = 2^5 3^8 5^3 7^2 11^3 13^8 17^6 19^8 23^4 29^4.$$

(The coding map π is not very efficient!). Certainly π is an injection.

The question which sentences of $\mathscr{L}$ are theorems of ZFC now has a simple formulation. It is known that there is a polynomial P_{ZFC} in finitely many variables, say in k variables, with coefficients in $\mathbb{Z}$, such that, for each sentence ϕ, $\phi \in \text{Thm(ZFC)}$ if and only if $\ulcorner\phi\urcorner$ is in the range of P_{ZFC} on $\mathbb{N}^k$. The existence of this polynomial arises from Matijacevič's solution to Hilbert's tenth problem : it is not possible to find an algorithm for testing an arbitrary polynomial equation $P(X_1, \ldots, X_n) = 0$ with coefficients in $\mathbb{Z}$ for the possession of a solution in integers. One can take k to be nine, and P_{ZFC} could be written down explicitly if one had sufficient fortitude. Thus, ZFC is consistent if and only if $\ulcorner \exists x_1(\neg(x_1 \hat{=} x_1)) \urcorner$ is not in the range of P_{ZFC}, and it is tempting to attempt to prove that this is the case. But Gödel's second incompleteness theorem shows that the attempt would be futile.

The question of the independence of sentences also has a simple formulation : a sentence ϕ is independent of

ZFC if and only if neither $\ulcorner\phi\urcorner$ nor $\ulcorner\neg\phi\urcorner$ is in the range of P_{ZFC}.

In fact, this number-theoretic approach has not been successful in proving *new* independence results, and success does not seem to be imminent.

The second approach is algebraic, and involves the building of models.

We have already given an example of this approach when we showed in Example 4.15 that the theory of dense total orders without endpoints is complete. In this example, we used Gödel's completeness theorem.

The point of this approach is that $ZFC + \phi$ is consistent if and only if $ZFC + \phi$ has a model, and one tries to build a model of $ZFC + \phi$, given a model of ZFC. Gödel was the first to be successful in this approach when he proved (in 1938) that Con ZFC implies Con(ZFC + CH). In fact, we have the following theorem.

4.19 THEOREM

Assume that ZFC has a model $\mathfrak{N}$. Then there is a submodel $\mathfrak{M}$ of $\mathfrak{N}$ such that $\mathfrak{M} \models ZFC + CH$.

(We shall prove in 7.38 that, if ZFC has a model $\mathfrak{N}$, then there is an extension $\mathfrak{M}$ of $\mathfrak{N}$ such that $\mathfrak{M} \models ZFC + CH$.)

In essence, Gödel discovered an effective method for building new models of ZFC by constructing submodels of a given model. What remained undiscovered for 25 years was an effective method for building extensions of a given model. The technique of forcing was discovered by Cohen in 1963 : forcing does give a method for building extensions, and it is the only such non-trivial method known today. Cohen's method allows one to extend a model by adding new elements in a controlled manner: the sentences which are true in the extended model are exactly those that are forced to be true. Forcing will be explained in Chapter 7.

4.20 NOTES

We have, in the formal definition of $\mathscr{L}(\hat{\in},\hat{=})$, taken the alphabet to consist of a set of positive integers. This is merely to avoid some metamathematical considerations, and is not important.

For a general background in mathematical logic, see [23] and [62], for example. For model theory, see Bell and Slomson ([7]), and Chang and Keisler ([10]), for example.

Gödel's completeness theorem 4.8 is [3, 4.9], [10, 1.3.21], and [23, §2.5]. The theorem applies to more general languages than $\mathscr{L}$ in the form that a theory is consistent if and only if it has a model.

Lists of the axioms of ZFC are given in [13], [44] and [49]. For a discussion of the Axiom of Choice, including many other formulations, see [43]. There is an interesting historical account of the origins of, and the controversies surrounding, this axiom in [53].

The Lindenbaum algebra is discussed in [7, Chapter 3, §4] and [44, §17].

For a discussion of Gödel's incompleteness theorems, see Smorynski [66] and the standard texts.

The formulation of NDH is due to Solovay.

An account of Matijacevič's work and of later results is given by Davis in [19]; see also [20].

We make one final remark. As we stated on page 64, we shall use Gödel's completeness theorem to step from an assertion about the existence of models to the fact that a sentence is relatively consistent with ZFC. In fact, because of the technical details of the method by which our models are constructed, it is possible in the cases which we consider to avoid the use of this theorem.

5 MARTIN'S AXIOM

In this chapter, we shall introduce Martin's Axiom (MA). We have three motives: first, a key step in our proof of the independence of NDH, Theorem 6.25, is given in the theory ZFC + MA + ¬CH; second, the techniques used here in the applications of MA are similar to those to be used later in the study of forcing; and third, we believe that some of the applications of MA to analysis that are given will be of general interest.

The axiom MA was first studied (as "Axiom A") by Martin and Solovay in [52]. The purpose of Martin and Solovay was to settle questions in ZFC + ¬CH that could not otherwise be resolved. In this aim, they were eminently successful: MA turned out to be an interesting and powerful axiom that does resolve many questions in ZFC + ¬CH, and indeed it is only rather recently that some combinatorial statements that are independent of MA + ¬CH have been discovered. At the present time, the study of some generalizations of MA is playing an important rôle in set theory.

We shall prove in Chapter 8 that the theory ZFC + MA + ¬CH is relatively consistent with ZFC.

Before dealing with MA, we shall first review the elementary theory of ordinals and cardinals that we require, and we shall formally state the Continuum Hypothesis. The proofs that we omit of "easily checked" results can all be found in many texts, for example in [49, Chapter 1].

We shall then introduce MA in the form of a version of the Baire category theorem for compact spaces, and we shall show that it is equivalent to MA', the original

Axiom A. Finally, we shall give some applications of MA in analysis: we hope these applications, all of which are well known, will give the reader some feeling how the axiom is used. However, it should be stressed that none of these applications will be used in the proofs of our main theorems, and so they could be omitted by a single-minded reader.

5.1 DEFINITION

Let $(P,<)$ be a totally ordered set.

(i) A subset Q of P is an <u>interval</u> if $c \in Q$ whenever $a,b \in Q$ and $c \in P$ with $a < c < b$.

(ii) A subset Q of P is an <u>initial segment</u> if $b \in Q$ whenever $a \in Q$ and $b \in P$ with $b < a$.

(iii) The ordering $<$ is a <u>well-ordering</u>, and $(P,<)$ is <u>well-ordered</u>, if each non-empty subset of P contains a minimum element.

Let P be a totally ordered set. For $a,b \in P$ with $a \leqslant b$, set

$$(a,b) = \{c \in P : a < c < b\},$$

$$[a,b] = \{c \in P : a \leqslant c \leqslant b\}.$$

Then (a,b) and $[a,b]$ are both intervals of P.

5.2 DEFINITION

Let $(P,<)$ and $(Q,<)$ be totally ordered sets. Then $(P,<) \prec (Q,<)$ if there is an order-isomorphism of $(P,<)$ onto an initial segment of $(Q,<)$.

It is easily checked that well-ordered sets have the following pleasant properties. Let $(P,<)$ and $(Q,<)$ be well-ordered sets. Then:

(i) either $(P,<) \prec (Q,<)$ or $(Q,<) \prec (P,<)$;

(ii) the only order-isomorphism of $(P,<)$ onto itself is the identity map.
It follows that, for well-ordered sets, an order-isomorphism specified in Definition 5.2 is unique, and that we can compare any two well-ordered sets: either they are order-isomorphic, or one is strictly "shorter" than the other, being order-isomorphic to a proper initial segment of the other.

5.3 DEFINITION

Let S be a set. Then S is <u>transitive</u> if each element of S is a subset of S, and S is an <u>ordinal</u> if S is transitive and if $(S,\in)$ is a well-ordered set.

Let us be more explicit about the condition that S be an ordinal. The relation $\in$ is a binary relation on each set S. For $\in$ to be a strict partial order on S, we require only that $a \in c$ whenever $a \in b$ and $b \in c$; the condition that, for each $a \in S$, $a \notin a$ is automatically true by the Axioms of Regularity and Comprehension. The condition that $\in$ be a total order does not follow from the condition that $\in$ be a strict partial order: for example, consider the set $S = \{\{1\},\{2\}\}$. However, it can be shown that the following are equivalent if S is a transitive set:

(i) $\in$ is a strict partial order on S;

(ii) $\in$ is a total order on S;

(iii) $\in$ is a well-ordering on S.

Thus, an ordinal is a set S satisfying the conditions: (i) if $a \in S$, then $a \subset S$; (ii) if $a \in b$ and $b \in c$, then $a \in c$ $(a,b,c \in S)$.

We shall usually denote ordinals by greek letters $\alpha,\beta,\gamma,\ldots$. Set $\alpha \leqslant \beta$ if $(\alpha,\in) \preccurlyeq (\beta,\in)$. Then, for ordinals α and β, either $\alpha \leqslant \beta$ or $\beta \leqslant \alpha$. It is easily checked that $\alpha < \beta$ if and only if $\alpha \in \beta$, and that $\alpha = \{\beta:\beta < \alpha\} = [0,\alpha)$. Thus, for example, "a finite subset of

$[0,\alpha)$" is the same as "a finite subset of α", and we shall later move freely between both designations.

The first three ordinals are $\emptyset$, $\{\emptyset\}$, and $\{\emptyset,\{\emptyset\}\}$; it is conventional to take these to be the numbers 0,1, and 2, respectively.

Let α be an ordinal and let S be a non-empty set. We shall often refer to maps from α to S. Such a map will be termed a <u>sequence of length α</u> (in S), and will be written

$$f = \langle x_\beta : \beta < \alpha \rangle;$$

the length of f is denoted by $\ell(f)$. If $\ell(f) = \alpha$ and if $\beta \leq \alpha$, then

$$f|\beta = \langle x_\gamma : \gamma < \beta \rangle.$$

Let f and g be sequences with $\ell(g) \leq \ell(f)$. Then f <u>extends</u> g if $f|\ell(g) = g$. An element of S^k will often be regarded as a sequence of length k in S.

Let $(P,<)$ be a well-ordered set. Then there is a unique ordinal α such that $(P,<)$ is order-isomorphic to $(\alpha,\in)$: this is a consequence of the Axiom of Replacement.

It is well known that the following statements are equivalent in the theory ZF:

(i) the Axiom of Choice;

(ii) the well-ordering principle, that each set can be well-ordered;

(iii) Zorn's Lemma.

Thus in the theory ZFC, the well-ordering principle and Zorn's Lemma hold. The following reformulation of the Axiom of Choice will be used in Chapter 7.

5.4 PROPOSITION

In ZF, AC is equivalent to the statement: for each set S, there is a function whose domain is an ordinal and whose range contains S.

Proof

If S is a set that can be well-ordered, then there is a bijection from an ordinal onto S. Conversely, if α is an ordinal and $\pi : \alpha \to S$ is a surjection, then S is well-ordered by the relation $<$, where $s < t$ in S if $\min\{\beta < \alpha : \pi(\beta) = s\} < \min\{\beta < \alpha : \pi(\beta) = t\}$. ∎

The collection of all sets having a given property is the class of sets with the property. Each set is a class, but consideration of Russell's paradox shows that, for example, the class of all sets is not a set. Let Ord be the class of all ordinals. Then Ord is not a set, for if S is a set of ordinals and if $T = \bigcup\{s : s \in S\}$, then $T \cup \{T\}$ is an ordinal which does not belong to S. Nevertheless, we shall apply the various notions of ordering, defined so far for sets, to the class Ord. Thus $(\text{Ord}, \in)$ is a well-ordered class.

Let α be an ordinal. Then $\alpha \cup \{\alpha\}$ is an ordinal, and it is the minimum element of the class $\{\beta \in \text{Ord} : \alpha < \beta\} : \alpha \cup \{\alpha\}$ is the successor of α, and it is written $\alpha + 1$. An ordinal α is a limit ordinal if $\alpha \neq 0$ and if α is not the successor of any ordinal.

Each set S of ordinals has a supremum in $(\text{Ord}, \in)$; it is $\bigcup S$, and it is denoted by $\sup S$.

We now introduce a notion which may be a novelty for analysts: we use the ordinals as an indexing set in a process (invented by Zermelo) that generates the class, or universe, of all sets. The universe grows from the simplest possible seed, the empty set, by repeated application of the power-set operation. The definition is by recursion on Ord.

5.5 DEFINITION

The cumulative hierarchy of sets is $\{V_\alpha : \alpha \in \mathrm{Ord}\}$, where:

$$V_o = \emptyset; \quad V_{\alpha+1} = P(V_\alpha);$$

$$V_\alpha = \cup\{V_\beta : \beta < \alpha\} \text{ for a limit ordinal } \alpha.$$

It is routine to prove by induction on α that each V_α is a transitive set, and so $V_\alpha \subset V_{\alpha+1}$ $(\alpha \in \mathrm{Ord})$. Thus $\langle V_\alpha : \alpha \in \mathrm{Ord}\rangle$ is an increasing sequence of sets.

Set

$$V = \cup\{V_\alpha : \alpha \in \mathrm{Ord}\}.$$

Then V is the universe of sets.

We need to know that every set appears somewhere in the above hierarchy, so that V is indeed the class of all sets. In fact, in the presence of the other axioms of ZF, this statement is equivalent to the Axiom of Regularity.

Let S be a set. The rank of S, denoted by $\mathrm{rk}(S)$, is the minimum ordinal α such that $S \in V_{\alpha+1}$. Intuitively, the rank of a set is a measure of its complexity, for it specifies the number of applications of the power-set operation needed to build the set.

The universe V is introduced here mainly as a preliminary to more sophisticated "Boolean-valued universes", to be defined in Chapter 7.

Let S and T be sets. Then S and T are equipotent if there is a bijection from S onto T; we write $S \sim T$, and say that S and T have the same cardinality. The relation $\sim$ is an equivalence relation on the class of all sets, and it is tempting to define the cardinal of a set to be the equivalence class to which it belongs. However, these equivalence classes are not sets, and this causes some problems. The now-standard method of bypassing these difficulties is due to von Neumann: we use ordinals and the Axiom of Choice to canonically define a representative of each equivalence class.

5.6 DEFINITION

Let α be an ordinal. Then α is a cardinal if, for each ordinal β with $\beta < \alpha$, α and β are not equipotent.

We often denote cardinals by the letters κ and λ.

Let S be a set. Then there is an ordinal which is equipotent to S, and so there is a minimum ordinal which is equipotent to S: this ordinal is the cardinal of S, and is denoted by $|S|$. Clearly, $|S|$ is a cardinal, and $|S| = |T|$ if and only if the sets S and T are equipotent.

The ordered set $(\mathbb{Z}^+,<)$ is order-isomorphic to an initial segment of $(\mathrm{Ord},\in)$. The ordinals in this set are the finite ordinals, and each is a cardinal; all other ordinals are infinite ordinals. The Axiom of Infinity guarantees that there is an infinite ordinal. The minimum infinite ordinal is denoted by ω; clearly, ω is a cardinal, and as such it is denoted by $\aleph_0$. Each infinite cardinal is a limit ordinal.

A set S is countable if $|S| \leqslant \aleph_0$, and S is uncountable if $|S| > \aleph_0$. A famous theorem of Cantor asserts that $|S| < |P(S)|$ for each set S, and so $\aleph_0 = |\mathbb{N}| < |P(\mathbb{N})| = |\mathbb{R}|$. In particular, there is an uncountable set. The minimum uncountable ordinal is ω_1; it is a cardinal, and as such it is denoted by $\aleph_1$.

A countable union of countable sets is countable. This fact will often be used in the form that, if $\{A_n\}$ is a countable family of sets such that $\bigcup A_n$ is uncountable, then at least one set A_n is uncountable.

Let κ and λ be non-zero cardinals, and let S and T be disjoint sets with $|S| = \kappa$ and $|T| = \lambda$. Then we define $\kappa + \lambda = |S \cup T|$ and $\kappa\lambda = |S \times T|$. Infinite cardinal arithmetic is essentially trivial: if κ and λ

are infinite cardinals, then $\kappa + \lambda = \kappa\lambda = \max\{\kappa, \lambda\}$.

Let κ be a cardinal. Then the minimum cardinal strictly greater than κ is denoted by κ^+. For example, $\aleph_1 = \aleph_0^+$. We set $\aleph_2 = \aleph_1^+$, etc.; in general, we define $\aleph_\alpha$ recursively for $\alpha \in \underset{\sim}{\text{Ord}}$ by setting $\aleph_\alpha = \aleph_\beta^+$ when $\alpha = \beta + 1$, and $\aleph_\alpha = \sup\{\aleph_\beta : \beta < \alpha\}$ when α is a limit ordinal. In each case, $\aleph_\alpha$ is a cardinal; it is denoted by ω_α when it is regarded as an ordinal.

The class $\{\aleph_\alpha : \alpha \in \underset{\sim}{\text{Ord}}\}$ is clearly the class of all cardinals, and so, for each set S, there is a unique ordinal α such that $|S| = \aleph_\alpha$. The problem is that of determining the α.

Let α be an ordinal. A subset S of α is <u>cofinal</u> in α if, for each $\beta < \alpha$, there exists $\gamma \in S$ with $\beta \leqslant \gamma$. The <u>cofinality</u> of α, denoted by $\text{cf}\,\alpha$ is the minimum ordinal β such that there is a function $f : \beta \to \alpha$ with $f(\beta)$ cofinal in α; an ordinal α is <u>regular</u> if $\text{cf}\,\alpha = \alpha$, and <u>singular</u> if $\text{cf}\,\alpha < \alpha$. Clearly each regular ordinal is a cardinal. For example, each ω_n is regular, but ω_ω is not regular because $\text{cf}\,\omega_\omega = \omega$.

We *claim* that, for each ordinal α, $\text{cf}\,\alpha$ is a regular cardinal. For let $\beta = \text{cf}\,\alpha$, and take $f : \beta \to \alpha$ with $f(\beta)$ cofinal in α. Define $g(\gamma) = \sup f([0,\gamma))$ $(\gamma < \beta)$. For each $\gamma < \beta$, $f([0,\gamma))$ is not cofinal in α, and so $g(\gamma) < \alpha$. If S is cofinal in β, then $g(S)$ is cofinal in α, and so $\text{cf}\,\beta = \text{cf}\,\alpha = \beta$. Thus β is a regular cardinal, as claimed.

Let κ be a cardinal. We next *claim* that it is a consequence of the Axiom of Choice that κ^+ is a regular cardinal. For let S be a cofinal subset of κ^+, and suppose, if possible, that $|S| \leqslant \kappa$. By AC, there is a function $F : S \to V$ such that, for each $\alpha \in S$, $F(\alpha) : \kappa \to \alpha$ is a surjection. Define $f : \kappa \times S \to \kappa^+$ by $f(\alpha,\beta) = F(\beta)(\alpha)$. Then f is a surjection since S is cofinal in κ^+. But $|\kappa \times S| = \kappa.|S| = \kappa$, a contradiction. So $|S| = \kappa^+$, and κ^+ is regular, as claimed.

The above claim will be used in the following form:

if $S \subset \kappa^+$ and if $|S| \leqslant \kappa$, then $\sup S < \kappa^+$. In fact, we shall almost always use only the easiest non-trivial form of this result: if S is a countable subset of ω_1, then $\sup S < \omega_1$. Here is an example of this application.

5.7 PROPOSITION

Let $f : \omega_1 \to \omega_1$ be a function. Then there is a limit ordinal η with $\eta < \omega_1$ such that $f(\alpha) < \eta$ $(\alpha < \eta)$.

<u>Proof</u>

Set $\alpha_1 = 1$ and inductively choose $\alpha_n < \omega_1$ so that $\alpha_{n+1} > \alpha_n$ and $\alpha_{n+1} \geqslant f(\alpha)$ $(\alpha < \alpha_n)$. Take $\eta = \sup \alpha_n$. ∎

We now come to cardinal exponentiation, which is not at all trivial.

Let κ be a cardinal. Then 2^κ is the cardinal of the set $\mathcal{P}(\kappa)$. Since $2^{\aleph_0} = |\mathbb{R}|$, $2^{\aleph_0}$ is termed the <u>cardinal of the continuum</u>, and is sometimes denoted by $\mathfrak{c}$. More generally, we have the following definition.

5.8 DEFINITION

Let κ and λ be cardinals. Then κ^λ is the cardinal of the set of functions from λ to κ.

It is easily checked that, if κ, λ, and μ are cardinals, then $(\kappa^\lambda)^\mu = \kappa^{\lambda\mu}$.

It remains to determine the value of 2^κ, and in particular of $2^{\aleph_0}$. Certainly $2^{\aleph_0} = \aleph_\alpha$ for some ordinal α. Cantor's theorem shows that, for each cardinal κ, $2^\kappa > \kappa$, and so $2^\kappa \geqslant \kappa^+$. In particular, $2^{\aleph_\alpha} \geqslant \aleph_{\alpha+1}$ for each ordinal α. Thus we have

$$2^{\aleph_0} \leqslant \aleph_0^{\aleph_0} \leqslant \aleph_1^{\aleph_0} \leqslant (2^{\aleph_0})^{\aleph_0} = 2^{\aleph_0}. \qquad (1)$$

A further constraint on the value of $2^{\aleph_0}$ is given in 5.9 (iii), below.

5.9 PROPOSITION

(i) Let κ be an infinite cardinal, and let λ be a cardinal. If $\mathrm{cf}\,\kappa \leq \lambda$, then $\kappa < \kappa^{\lambda}$.

(ii) If $|A_n| < 2^{\aleph_o}$ $(n \in \mathbb{N})$, then $|\cup A_n| < 2^{\aleph_o}$.

(iii) $\mathrm{cf}\,2^{\aleph_o} > \aleph_o$, and so $2^{\aleph_o} \neq \aleph_{\omega}$.

Proof

(i) Let $f : \lambda \to \kappa$ be a function with cofinal range, and let H be a map from κ into the set of functions from λ into κ. Define $h : \lambda \to \kappa$ by requiring that $h(\alpha)$ be the minimum ordinal in the set

$$\kappa \setminus \{H(\beta)(\alpha) : \beta < f(\alpha)\}$$

for $\alpha < \lambda$. Then h is not in the range of H, and so H is not surjective. Thus $\kappa < \kappa^{\lambda}$.

(ii) Set $\kappa = \sup\{|A_n|\}$. Then $\kappa \leq \aleph_o . 2^{\aleph_o} = 2^{\aleph_o}$. We can suppose that each A_n is infinite and that $A_n \subset A_{n+1}$ $(n \in \mathbb{N})$. Then $\kappa \leq |\cup A_n| \leq \aleph_o . \kappa = \kappa$. If $\langle |A_n| : n \in \mathbb{N}\rangle$ is eventually constant, then certainly $\kappa < 2^{\aleph_o}$. If $\langle |A_n| : n \in \mathbb{N}\rangle$ is not eventually constant, then $\mathrm{cf}\,\kappa = \aleph_o$, and so, by (i), $\kappa < \kappa^{\aleph_o} \leq (2^{\aleph_o})^{\aleph_o} = 2^{\aleph_o}$.

(iii) Apply (i) with $\kappa = 2^{\aleph_o}$ and $\lambda = \aleph_o$: since $\kappa = \kappa^{\lambda}$, $\mathrm{cf}\,\kappa > \lambda$. ∎

5.10 DEFINITION

The <u>Continuum Hypothesis</u> (CH) is the sentence :

$$2^{\aleph_o} = \aleph_1 .$$

The <u>Generalized Continuum Hypothesis</u> (GCH) is the sentence :

$$2^{\aleph_\alpha} = \aleph_{\alpha+1} \quad \text{for each ordinal } \alpha.$$

It was, of course, Cantor, in the nineteenth century who first considered the Continuum Hypothesis.

There is an alternative formulation of CH which does not use the theory of ordinals: "if $S \subset P(\mathbb{N})$, then either S is equipotent to a subset of $\mathbb{N}$ or S is equipotent to $P(\mathbb{N})$."

We now give a calculation in cardinal arithmetic that uses CH. The result will be used in Chapter 8. Recall that results that are claimed in the theory ZFC + CH are labelled "CH".

5.11 PROPOSITION (CH)

$$\aleph_2^{\aleph_o} = \aleph_2.$$

Proof

For $\xi \leq \omega_2$, take S_ξ to be the set of functions from ω to ξ. Then $S_{\omega_2} = \bigcup\{S_\xi : \xi < \omega_2\}$. Now, for $\xi < \omega_2$, $|S_\xi| \leq \aleph_1^{\aleph_o}$; by (1), $|S_\xi| \leq 2^{\aleph_o^2}$, and, by CH, $|S_\xi| \leq \aleph_1$. Thus $\aleph_2^{\aleph_o} = |S_{\omega_2}| = \aleph_2.\aleph_1 = \aleph_2$. ∎

The Axiom of Choice is generally (but not universally) accepted, and we shall work in ZFC. Should we accept CH as an axiom? At a very naive level, CH is plausible if it is stated as "there are no sets S with $\aleph_o < |S| < \mathfrak{c}$", just because we cannot think of such a set S. Also, if we reject CH, we have a problem deciding why $2^{\aleph_o}$ should take any other particular value. On the other hand, CH seems improbable in the form "there is a well-ordering of $\mathbb{R}$ in which each element has only countably many predecessors". One of the tasks of set theory is to explore the consequences of the assumption of a particular value for $2^{\aleph_o}$, and to find other "natural" axioms which determine this value. It may well be that our understanding of sets will develop sufficiently for a concensus to emerge on the "most reasonable" value of $2^{\aleph_0}$.

The axioms of ZFC and the rôle of CH have been known (in some form) for 80 years. We now come to an axiom, Martin's Axiom, that goes beyond ZFC, which was formulated less than 20 years ago.

The force of CH is that, if S is a set with $|S| < 2^{\aleph_0}$, then S is "small", for then S must be countable. Martin's Axiom attempts to preserve the feature that, if $|S| < 2^{\aleph_0}$, then S is "small" in some sense, without resorting to the extreme position that each small set be countable. For example, MA implies that each subset S of $\mathbb{R}$ with $|S| < 2^{\aleph_0}$ is small, in the sense that S has Lebesgue measure zero.

Our route to MA is meant to be the most scenic one for analysts: it passes through the green pastures of Baire category theory.

Recall that a topological space X is a Baire space if, given a countable family $\{U_n\}$ of dense, open subsets of X, it follows that $\cap U_n$ is also dense in X. One form of the Baire category theorem is the following.

5.12 THEOREM

Each compact topological space is a Baire space. ∎

Informally, the theorem asserts that the intersection of a "small" set of dense, open subsets of a compact space is dense. A natural first attempt at a weaker axiom than CH is the following.

> Let X be a compact space, and let $\mathcal{U}$ be a family of dense, open sets in X with $|\mathcal{U}| < 2^{\aleph_0}$. Then $\cap\mathcal{U}$ is dense in X. (*)

However, (*) already implies CH. For consider

the space $W^* = [0,\omega_1]$, a compact space with respect to the order topology. Let $X = (W^*)^{\mathbb{N}}$. By Tychonoff's theorem, X is a compact space for the product topology. For $\sigma < \omega_1$, set

$$U_\sigma = \{f \in X : f(n) \in (\sigma,\omega_1) \text{ for some } n \in \mathbb{N}\}.$$

Then U_σ is dense and open in X, and $|\{U_\sigma : \sigma < \omega_1\}| = \aleph_1$. Take $f \in X$. Since $f(\mathbb{N}) \cap [0,\omega_1)$ is countable, there exists $\sigma < \omega_1$ with $f(\mathbb{N}) \subset [0,\sigma] \cup \{\omega_1\}$. Clearly $f \notin U_\sigma$, and so $\bigcap\{U_\sigma : \sigma < \omega_1\} = \emptyset$. Thus, if (*) holds, then necessarily $\aleph_1 = 2^{\aleph_0}$.

We actually obtain MA by modifying (*): we restrict the class of compact spaces that are considered.

5.13 DEFINITION

Let X be a Hausdorff topological space. Then X satisfies the countable chain condition if each family consisting of pairwise disjoint, non-empty, open subsets of X is countable.

We also say that a topological space which satisfies the countable chain condition is ccc.

A topological space is separable if it has a countable, dense subset. Clearly, each separable space is ccc, but the condition of being ccc is strictly weaker than that of being separable: we shall show in 6.2 that the compact space $\{0,1\}^\kappa$ is ccc for each cardinal κ, but it is clear that $\{0,1\}^\kappa$ is not separable if $\kappa > 2^{\aleph_0}$.

5.14 DEFINITION

Martin's Axiom (MA) is the sentence:

Let X *be a compact topological space satisfying the countable chain condition, and let* $\mathcal{U}$ *be a family of dense, open sets in* X

with $|\mathcal{U}| < 2^{\aleph_0}$. *Then* $\bigcap\{U : U \in \mathcal{U}\}$ *is dense in* X.

Certainly, CH implies MA. We shall see in due course that, if ZFC is consistent, then so is the theory ZFC + MA + ¬CH, and so MA does not imply CH. First, however, we give a sentence which is equivalent to MA in ZFC. It is less natural, but easier to use, than the actual statement.

5.15 DEFINITION

Let $(P,<)$ be a partially ordered set.

(i) A subset Q of P is an <u>antichain</u> if, for each $a,b \in Q$ with $a \neq b$, a and b are incompatible.

(ii) P satisfies the <u>countable chain condition (ccc)</u> if each antichain in P is countable; $(P,<)$ is then a <u>ccc partial order</u>.

(iii) Let $\mathcal{D}$ be a subset of $\mathcal{P}(P)$. A subset G of P is <u>$\mathcal{D}$-generic</u> if $G \cap D \neq \emptyset$ $(D \in \mathcal{D})$.

A Boolean algebra B <u>is ccc</u> if the partially ordered set $(B\backslash\{0\},<_B)$ (see page 27) is ccc. Thus B is ccc if each subset S of $B\backslash\{0\}$ such that $a \wedge b = 0$ whenever a and b are distinct elements of S is countable.

Note that, if $f = \langle a_\alpha : \alpha < \omega_1\rangle$ is a sequence of length ω_1 in a ccc Boolean algebra such that $a_\alpha \leq a_\beta$ $(\alpha \leq \beta < \omega_1)$, then f is eventually constant.

5.16 LEMMA

Let $(P,<)$ be a ccc partial order, and let (B,π) be the completion of P. Then B is ccc. Further, if $|P| \leq 2^{\aleph_0}$, then $|B| \leq 2^{\aleph_0}$.

<u>Proof</u>

Since $\pi(P)$ is dense in $B\backslash\{0\}$, B is ccc, and, for each $b \in B\backslash\{0\}$, there is a set $A \subset \pi(P)$ such that $b = \bigvee A$. Since B is ccc, A can be taken to be countable. Now suppose that $|P| \leqslant 2^{\aleph_0}$. Then $|B| \leqslant (2^{\aleph_0})^{\aleph_0} = 2^{\aleph_0}$. ∎

5.17 DEFINITION

MA' is the sentence:

Let P *be a partially ordered set satisfying the countable chain condition, and let* $\mathcal{D}$ *be a family of dense subsets of* P *with* $|\mathcal{D}| < 2^{\aleph_0}$. *Then there is a* $\mathcal{D}$*-generic filter in* P.

5.18 LEMMA

Suppose that MA' holds when restricted to partially ordered sets P with $|P| < 2^{\aleph_0}$. Then MA' holds.

Proof

Let P and $\mathcal{D}$ be as in the statement MA'. We define by induction certain subsets Q_n of P. Form Q_o by choosing one element of D for each $D \in \mathcal{D}$. Now suppose that Q_n has been constructed. For each pair $\{a,b\}$ of compatible (not necessarily distinct) elements of Q_n and each $D \in \mathcal{D}$, choose $c \in D$ with $c \leqslant a$ and $c \leqslant b$: such an element c always exists. Form Q_{n+1} by adjoining to Q_n all these specified elements c of D. Set $Q = \bigcup Q_n$. Then $|Q_n| < 2^{\aleph_0}$ for each n, and so, by 5.9(ii), $|Q| < 2^{\aleph_0}$.

For each $D \in \mathcal{D}$, $D \cap Q$ is dense in Q. If $a,b \in Q$, then a and b are compatible in Q if they are compatible in P, and so Q is ccc. By hypothesis, there is a filter G in Q such that $G \cap D \cap Q \neq \emptyset$ $(D \in \mathcal{D})$.

As a subset of P, G is a prefilter, and so G is contained in a filter, say H, in P. Clearly H is $\mathcal{D}$-generic. ∎

5.19 LEMMA

Suppose that MA' holds when restricted to complete Boolean algebras. Then MA' holds.

<u>Proof</u>

Let P and $\mathcal{D}$ be as in the statement MA'. By 5.18, we may suppose that $|P| < 2^{\aleph_0}$.

For $a \in P$, set $D_a = \{b \in P : b < a \text{ or } b \perp a\}$. Note that, if $b \in D_a$ and $c \leqslant b$, then $c \in D_a$. Clearly, each D_a is dense in P, and so $D_a \cap D_b$ is dense for $a,b \in P$. Thus we may suppose that $D_a \cap D_b \in \mathcal{D}$ for $a,b \in P$.

Let (B,π) be a completion of P, so that B is ccc. Let $D \in \mathcal{D}$, and take $b \in B\backslash\{0\}$. By 2.9(i), there exists $a \in P$ with $\pi(a) \leqslant b$, and there exists $d \in D$ with $d \leqslant a$. By 2.9(ii), $\pi(d) \leqslant b$, and so $\pi(D)$ is dense in $B\backslash\{0\}$. By hypothesis, there is a filter G in B with $G \cap \pi(D) \neq \emptyset$ $(D \in \mathcal{D})$.

Let $H = \pi^{-1}(G)$. Then $H \cap D \neq \emptyset$ $(D \in \mathcal{D})$. Take $a,b \in H$. Since $D_a \cap D_b \in \mathcal{D}$, there exists $c \in H \cap D_a \cap D_b$. Since G is a filter, $\pi(c) \not\perp \pi(a)$, and so, by 2.9(iii), $c \not\perp a$. Thus $c \leqslant a$, and, similarly, $c \leqslant b$. Thus H is a prefilter in P, and hence, by 2.9(ii), H is a filter. ∎

5.20 THEOREM

The statements MA and MA' are equivalent.

<u>Proof</u>

Suppose that MA holds. To establish MA', it suffices to establish it for complete Boolean algebras.

Let B be a complete ccc Boolean algebra, and let $\mathcal{D}$ be a family of dense subsets of B with $|\mathcal{D}| < 2^{\aleph_0}$. Let $S(B)$ be the Stone space of B, so that $S(B)$ is a compact space, and let $\pi : B \to P(S(B))$ be the associated map (as in equation (4) of page 33). For $D \in \mathcal{D}$, set $\tilde{D} = \bigcup\{\pi(d) : d \in D\}$. By 2.13(iii), each $\tilde{D}$ is a dense, open set in $S(B)$. Since B is ccc, $S(B)$ is ccc, and so, by

MA, there exists $G \in \bigcap\{\tilde{D} : D \in \mathcal{D}\}$. Now G is an ultrafilter in B, and $G \cap D \neq \emptyset$ $(D \in \mathcal{D})$. Thus G is a $\mathcal{D}$-generic filter in B, and so MA' holds.

Conversely, suppose that MA' holds. Let X be a compact space satisfying ccc, and let $\mathcal{U}$ be a family of dense, open subsets of X with $|\mathcal{U}| < 2^{\aleph_0}$. Let $B = \mathcal{R}(X)$, the regular-open algebra of X (Example 2.7). Note that, since X is normal, for each non-empty, open set V in X, there exists $a \in B\backslash\{0\}$ with $\bar{a} \subset V$. Thus, to show that $\bigcap\mathcal{U}$ is dense in X, it suffices to show that we have $\bigcap\{U \cap \bar{a} : U \in \mathcal{U}\} \neq \emptyset$ for each $a \in B\backslash\{0\}$.

Take $a_o \in B\backslash\{0\}$, and let $P = \{b : b \leqslant a_o\}$. Clearly, P is ccc. For $U \in \mathcal{U}$, set $D_U = \{a \in P : \bar{a} \subset U\}$. Since U is dense and open in X, for each $a \in P$ there exists $b \in D_U$ with $b \leqslant a$, and so D_U is dense in P.

By MA', there is a filter G in P with $G \cap D_U \neq \emptyset$ $(U \in \mathcal{U})$, and so, for each $U \in \mathcal{U}$, there exists $b \in G$ with $\bar{b} \subset U \cap \bar{a}_o$. Since $\{\bar{b} : b \in G\}$ has the finite intersection property and X is compact, $\bigcap\{\bar{b} : b \in G\} \neq \emptyset$, and so $\bigcap\{U \cap \bar{a}_o : U \in \mathcal{U}\} \neq \emptyset$, as required. Thus MA holds. ∎

It is a surprising aspect of the nature of things that the simple generalization of the Baire category theorem given by MA produces a powerful axiom. Our applications illustrate how MA can be exploited.

Theorems which hold in the theories ZFC + MA and ZFC + MA + ¬CH are labelled "(MA)" and "(MA + ¬CH)", respectively.

5.21 DEFINITION

A <u>rational interval</u> of $\mathbb{R}$ is a subset of $\mathbb{R}$ of the form (x,y), where $x,y \in \mathbb{Q}$ and $x < y$. The set $\mathcal{I}$ is the family of open subsets of $\mathbb{R}$, each of which is a finite union of rational intervals.

Certainly, the set $\mathcal{I}$ is countable.

In the next theorem, μ is Lebesgue measure on $\mathbb{R}$.

5.22 THEOREM (MA)

Let $\mathcal{F}$ be a family of measurable subsets of $\mathbb{R}$ such that $\mu(F) = 0$ $(F \in \mathcal{F})$ and $|\mathcal{F}| < 2^{\aleph_0}$. Let $S = \bigcup\{F : F \in \mathcal{F}\}$. Then S is measurable, and $\mu(S) = 0$.

<u>Proof</u>

Fix $\varepsilon > 0$, and let

$$P_\varepsilon = \{a \subset \mathbb{R} : a \text{ is open, } \mu(a) < \varepsilon\}.$$

For $a,b \subset P_\varepsilon$, set $a \leqslant b$ if $b \subset a$. Then $(P_\varepsilon, <)$ is a partially ordered set.

Suppose, if possible, that C is an uncountable antichain in P_ε. For $n \in \mathbb{N}$, let

$$C_n = \{c \in C : \mu(a) < (1 - \tfrac{1}{n})\varepsilon\}.$$

Then there exists $m \in \mathbb{N}$ such that C_m is uncountable. For each $a \in C_m$, choose $U_a \in I$ with $U_a \subset a$ and $\mu(a - U_a) < \varepsilon/m$. Since I is countable and C_m is uncountable, there exist distinct elements $a,b \in C_m$ with $U_a = U_b$, and in this case we have

$$\begin{aligned}\mu(a \cup b) &\leqslant \mu(a) + \mu(b - U_a)\\ &< (1 - \tfrac{1}{m})\varepsilon + \tfrac{1}{m}\varepsilon = \varepsilon,\end{aligned}$$

and so $a \cup b \in P_\varepsilon$. But $a \cup b \leqslant a$ and $a \cup b \leqslant b$, a contradiction of the fact that $a \perp b$. Thus P_ε is ccc.

For $F \in \mathcal{F}$, let $D_F = \{b \in P_\varepsilon : F \subset b\}$. For each $a \in P_\varepsilon$ and each $F \in \mathcal{F}$, there is an open set $U \subset \mathbb{R}$ with $F \subset U$ and $\mu(U) < \varepsilon - \mu(a)$; we have $a \cup U \in D_F$ and $a \cup U \leqslant a$. Thus each D_F is dense in P_ε.

Since $|\mathcal{F}| < 2^{\aleph_0}$, MA implies that there is a filter G in P_ε with $G \cap D_F \neq \emptyset$ $(F \in F)$, and so, for each $F \in F$, there exists $b \in G$ with $F \subset b$. Set $V = \bigcup\{b : b \in G\}$, an open set in $\mathbb{R}$. For each $b \in G$ and each $x \in b$, there exists $c \in I$ with $x \in c \subset b$, and so there exists $(b_n) \subset G$ with $V = \bigcup b_n$. For $n \in \mathbb{N}$, take $c_n \in P_\varepsilon$ with $c_n \leqslant b_j$ $(j = 1,\ldots,n)$. Then, for each n, we have $\mu(\bigcup_{j=1}^{n} b_j) \leqslant \mu(c_n) < \varepsilon$, and so $\mu(V) \leqslant \varepsilon$.

For each $F \in F$, there exists $b \in G$ with $F \subset b$, and so $S \subset V$.

We have shown that, for each $\varepsilon > 0$, there is an open set V with $V \supset S$ and $\mu(V) \leqslant \varepsilon$. Hence $\mu(S) = 0$. ∎

Consider $(\mathbb{R}, <)$, where $<$ is the usual order. Then $\mathbb{R}$ is a totally ordered set which has neither a maximum nor a minimum element, and $\mathbb{R}$ is connected and separable with respect to the order topology, i.e., with respect to the usual topology. It is easy to see that these properties characterize $\mathbb{R}$, in the sense that each totally ordered set with these properties is order-isomorphic to $\mathbb{R}$. In 1920, Souslin raised the question whether or not "separable" could be replaced by "ccc" in this result. More generally, let $(P,<)$ be a totally ordered set which satisfies ccc in the order topology. Is P necessarily separable?

5.23 DEFINITION

A totally ordered set $(P,<)$ is a Souslin line if $(P,<)$ satisfies the countable chain condition, but is not separable in the order topology. Souslin's Hypothesis (SH) is the statement:

There are no Souslin lines.

We shall show that SH follows from MA + ¬CH. On the other hand, it can be shown (see [49,II], [21]) that ¬SH follows from a principle called "diamond", which is consistent with ZFC + CH. Thus SH is independent of ZFC.

5.24 THEOREM

If MA and ¬CH hold, then SH holds.

Proof

Suppose, if possible, that MA and ¬CH hold, and that SH is false, and let $(P,<)$ be a Souslin line.

Let M be a maximal family in $(\mathcal{P}(P), \subset)$ of pairwise disjoint, non-empty, open intervals of P, each of which is separable, and let $E = \cup\{I : I \in M\}$. Since P is ccc, M is countable, and so E is separable. Since P is not separable, E is not dense in P, and so there is a non-empty, open interval J in P with $J \cap E = \emptyset$. Now $(J,<)$ is a Souslin line, and so, replacing P by J, we may suppose that each non-empty, open interval in P is non-separable. This implies that each non-empty, open set in P is a countable union of pairwise disjoint, non-separable, open intervals: these intervals are the components of the open set.

We now construct a sequence $\langle U_\alpha : \alpha < \omega_1 \rangle$ of dense, open subsets of P such that $U_o = P$ and such that, for each α, β with $\alpha < \beta < \omega_1$:

(i) $U_\beta \subset U_\alpha$;

(ii) if I is a component of U_α, then $I \not\subset U_\beta$.

Given U_α, we form $U_{\alpha+1}$ by deleting one point from each component of U_α. Since each component of U_α is non-separable, the deleted point was not isolated in U_α, and so $U_{\alpha+1}$ is dense in P.

Now let β be a limit ordinal with $\beta < \omega_1$, and suppose that U_α has been defined for each $\alpha < \beta$. For each $\alpha < \beta$, there is a countable set S_α such that, if $a < b$ and a and b are in different components of U_α, then there exists $c \in S_\alpha$ with $a < c < b$. Set $S = \cup\{S_\alpha : \alpha < \beta\}$, a countable set. Take U_β to be the union of those intervals (x,y) of P such that $x < y$ and (x,y) is contained in a component of U_α for each $\alpha < \beta$. Then U_β is open, $U_\beta \subset U_\alpha$ $(\alpha < \beta)$, and, if I is a component of U_α, then $I \not\subset U_{\alpha+1}$, and so $I \not\subset U_\beta$. Finally, we show that U_β is dense in P. For let J be a non-empty, open interval in P. Since J is non-separable, there exists $(x,y) \subset J$ with $(x,y) \cap S = \emptyset$, and clearly $(x,y) \subset U_\beta$. Thus U_β is dense, and the construction continues.

Let Q_α be the family of components of U_α, let $Q = \bigcup\{Q_\alpha : \alpha < \omega_1\}$, and let $D_\alpha = \bigcup\{Q_\beta : \alpha < \beta < \omega_1\}$. For $I,J \in Q$, set $I \leqslant J$ if $I \subset J$. Then $(Q,<)$ is a partially ordered set; clearly, Q is ccc, and each D_α is dense in Q. By MA, there is a filter G in Q with $G \cap D_\alpha \neq \emptyset$ $(\alpha < \omega_1)$. For each β, let $J_\beta \in Q$ be such that $G \cap Q_\beta = \{J_\beta\}$. Then $J_\beta \subsetneq J_\alpha$ whenever $\alpha < \beta < \omega_1$, and so, for each $\beta < \omega_1$, there is a non-empty open interval $V_\beta \subset J_\beta \setminus J_{\beta+1}$. But then $\{V_\beta : \beta < \omega_1\}$ is an uncountable antichain in Q, a contradiction.

We conclude that SH holds. ∎

We now give an interesting combinatorial principle that has a number of applications.

5.25 THEOREM (MA)

Let F be a prefilter in $P(\mathbb{N})$ consisting of infinite sets, and suppose that $|F| < 2^{\aleph_0}$. Then there is an infinite set $\sigma \subset \mathbb{N}$ such that $\sigma\setminus\tau$ is finite for each $\tau \in F$.

<u>Proof</u>

Let

$$P = \{(a,b) : a \subset \mathbb{N}, \ a \text{ finite}, \ b \in F\},$$

and set $(a_1,b_1) \leqslant (a_2,b_2)$ in P if $a_2 \subset a_1$, if $b_1 \subset b_2$, and if $a_1\setminus a_2 \subset b_2$. Then $(P,<)$ is a partially ordered set.

Suppose that $(a,b_1), (a,b_2) \in P$. Since F is a prefilter, there exists $c \in F$ with $c \subset b_1$ and $c \subset b_2$, and then $(a,c) \leqslant (a,b_1)$ and $(a,c) \leqslant (a,b_2)$. Thus, if

$(a_1,b_1) \perp (a_2,b_2)$ in P, then $a_1 \neq a_2$. Since there are only countably many finite subsets of $\mathbb{N}$, P is ccc.

For $\tau \in F$, set $D_\tau = \{(a,b) \in P : b \subset \tau\}$. For each $(a,b) \in P$, $(a,b) \not\perp (a,\tau)$, and so each D_τ is dense in P. For $k \in \mathbb{N}$, set $E_k = \{(a,b) \in P : |a| \geq k\}$, and take $(a,b) \in P$. Since b is infinite, there exists $c \subset b$ with $|c| = k$; we have $(a \cup c, b) \in E_k$ and $(a \cup c, b) \leq (a,b)$, and so each E_k is dense in P.

By MA, there is a filter G in P with $G \cap D_\tau \neq \emptyset$ $(\tau \in F)$ and $G \cap E_k \neq \emptyset$ $(k \in \mathbb{N})$.

Set

$$\sigma = \bigcup\{a : (a,b) \in G \text{ for some } b \in F\}.$$

For each $k \in \mathbb{N}$, there exists $(a,b) \in G$ with $|a| \geq k$, and so σ is an infinite set. For each $\tau \in F$, there exists $(a_\tau, b_\tau) \in G$ with $b_\tau \subset \tau$. If possible, choose $m \in \sigma \backslash (a_\tau \cup \tau)$. Then there exists $(a,b) \in G$ with $m \in a$. Since G is a filter, there exists $(c,d) \in G$ with $(c,d) \leq (a,b)$ and $(c,d) \leq (a_\tau, b_\tau)$. We have $a \subset c$ and $c \backslash a_\tau \subset b_\tau \subset \tau$, and so $m \in c \subset a_\tau \cup \tau$, a contradiction. Thus $\sigma \backslash \tau \subset a_\tau$, a finite set.

This completes the proof of the theorem. ∎

5.26 THEOREM (MA)

Let X and Y be subsets of $\mathbb{R}$ with $X \subset Y$. Suppose that $|Y| < 2^{\aleph_0}$. Then X is a G_δ-set in Y.

<u>Proof</u>

Let I be the family of finite unions of rational intervals in $\mathbb{R}$, as in 5.21, say $I = \{I_n : n \in \mathbb{N}\}$, and let F be the family of finite subsets of Y. Set

$$\tau_S = \{k \in \mathbb{N} : I_k \cap S = X \cap S\} \quad (S \in F).$$

Since $\tau_{S \cup T} \subset \tau_S \cap \tau_T$ $(S, T \in F)$, $\{\tau_S : S \in F\}$ is a prefilter on $\mathbb{N}$. Further, each set τ_S is infinite.

By 5.25, there is an infinite set $\sigma \subset \mathbb{N}$ such that $\sigma \backslash \tau_S$ is finite for each $S \in F$. For $j \in \mathbb{N}$, set $U_j = \bigcup\{I_k : k > j,\ k \in \sigma\}$, and let $H = \bigcap\{U_j : j \in \mathbb{N}\}$. Each U_j is open, and so H is a G_δ-set in $\mathbb{R}$.

Let $a \in X$. Then $\tau_{\{a\}} = \{k \in \mathbb{N} : a \in I_k\}$. Take $j \in \mathbb{N}$. Since σ is infinite and $\sigma \backslash \tau_{\{a\}}$ is finite, there exists $k > j$ with $k \in \sigma$ and $a \in I_k$, and so $a \in U_j$.

Let $b \in Y \backslash X$. Then $\tau_{\{b\}} = \{k \in \mathbb{N} : b \notin I_k\}$. Since σ is infinite and $\sigma \backslash \tau_{\{b\}}$ is finite, there exists $j \in \mathbb{N}$ such that, for each $k > j$ with $k \in \sigma$, $b \notin I_k$, and so $b \notin U_j$.

It follows that $H \cap Y = X$, giving the result. ∎

5.27 COROLLARY (MA)

Let Y be an infinite subset of $\mathbb{R}$ such that $|Y| < 2^{\aleph_0}$. Then $|P(Y)| = 2^{\aleph_0}$.

<u>Proof</u>

Let G be the family of G_δ-subsets of $\mathbb{R}$. Then $|G| = 2^{\aleph_0}$, and, by 5.26, the map $H \mapsto H \cap Y$, $G \to P(Y)$, is a surjection. ∎

5.28 COROLLARY

(i) (MA + ¬CH) $2^{\aleph_1} = 2^{\aleph_0}$.

(ii) (CH) $2^{\aleph_1} > 2^{\aleph_0}$. ∎

Thus whether or not $2^{\aleph_1} = 2^{\aleph_0}$ cannot be determined in ZFC.

We have introduced MA as a form of the category theorem for compact spaces. We conclude the chapter by noting that MA implies a similar result for complete metric spaces.

5.29 THEOREM (MA)

Let X be a complete, ccc metric space, and let U be a family of dense, open sets in X with $|U| < 2^{\aleph_0}$. Then $\bigcap\{U : U \in U\}$ is dense in X.

<u>Proof</u>

Let βX be the Stone-Čech compactification of X; as before, we write $\bar{S}$ for the closure in βX of a subset S of X. Since X is ccc, βX is ccc, and so it suffices to show that X is a G_δ-subset of βX.

For each open set V in X, let $\tilde{V}$ be an open set in βX with $\tilde{V} \cap X = V$. For $n \in \mathbb{N}$, let $\mathcal{V}_n$ be the family of open balls of radius $1/n$ in X, and let $W_n = \bigcup\{\tilde{V} : V \in \mathcal{V}_n\}$. We *claim* that $X = \bigcap W_n$. Take $x \in \bigcap W_n$, and, for each n, choose $V_n \in \mathcal{V}_n$ with $x \in \tilde{V}_n$. Set $S = (\bigcap \bar{V}_n) \cap X$. Since X is complete, S is a singleton in X, say $S = \{y\}$. Suppose, if possible, that $y \neq x$. Then there is a neighbourhood V of y in X such that $x \notin \bar{V}$. But eventually $V_n \subset V$, and so $\tilde{V}_n \subset \tilde{V} \subset \bar{V}$, a contradiction. So $y = x$, $x \in X$, and $X = \bigcap W_n$, giving the result. ∎

5.30 NOTES

The theory of ordinals and cardinals that we have given is quite standard. Proposition 5.9 is a form of König's theorem ([49,I.10.40]).

Most of the results about MA that we have given were first proved by Martin and Solovay [52], and are included in [44] and [49]. The book [34] of Fremlin is a comprehensive account of MA. It includes our applications, and many more, often in more general form, and it has a detailed account of the historical sources. Two articles about MA are [57] and [63]; a more recent account is [70].

Generalizations of MA to larger classes of partially ordered sets have been studied. A certain maximal generalization is called MM ('Martin's Maximum'). See [32] and [45, Chapter 7]. It is known that, assuming the consistency of certain large cardinal axioms, MM + ¬CH is also consistent. Curiously, MM + ¬CH implies that $2^{\aleph_0} = \aleph_2$.

Another variant of MA is PFA (the "Proper Forcing Axiom"): see [45] and [61]. In fact, MM implies PFA and PFA implies NDH: even the definitions of MM and of PFA are beyond the scope of this book.

6 GAPS IN ORDERED SETS

Our main story about the existence of discontinuous homomorphisms from the algebras $C(X,\mathbb{C})$ into Banach algebras was left in Chapter 3. In the language of Chapter 4, the consistency of the axiom NDH with ZFC will be established by the construction of a model of ZFC + NDH given a model of ZFC. We take a major step towards this goal in the present chapter by proving in Theorem 6.25 that, with MA + $\neg$CH, NDH follows from the fact that there is no embedding of a certain space $(\mathfrak{R},<)$ into $(\mathbb{N}^{\mathbb{N}},<_F)$.

Our technique for this is to study gaps in ordered sets, and in particular in $(\mathbb{N}^{\mathbb{N}},<_F)$. Roughly, a (κ,λ)-gap in an ordered set is a pair of sequences

$$\langle f_\alpha, g_\beta : \alpha < \kappa, \beta < \lambda\rangle$$

such that $\langle f_\alpha\rangle$ is strictly increasing, $\langle g_\beta\rangle$ is strictly decreasing, and $f_\alpha < g_\beta$ for all α,β, but such that there is no h with $f_\beta < h < g_\beta$ for all α,β. It is a little exercise to show that there are no $(\aleph_0,\aleph_0)$-gaps in $(\mathbb{N}^{\mathbb{N}},<_F)$; it is a famous theorem of Hausdorff that there are $(\aleph_1,\aleph_1)$-gaps in $(P(\mathbb{N}),<_F)$, and the same result holds in $(\mathbb{N}^{\mathbb{N}},<_F)$.

The main result of this chapter requires Theorems 6.15 and 6.16 as preliminaries, and the earlier technical result 6.12 will also be seriously used in Chapter 8, but Theorem 6.9 will not be needed later.

Before beginning the main topic of this chapter, we give a well-known result, often called the <u>Δ-system lemma</u>, that will be required several times.

6.1 PROPOSITION

Let F be an uncountable family of finite subsets of a set X. Then there is a finite subset T of X and an uncountable subfamily G of F such that $A \cap B = T$ whenever A and B are distinct elements of G.

Proof

We may suppose that $|F| = \aleph_1$: in this case, $|\bigcup F| = \aleph_1$, and so we may suppose that $X = \omega_1$.

For each $k \in \mathbb{N}$, set $F_k = \{\sigma \in F : |\sigma| = k\}$. Then $F = \bigcup\{F_k : k \in \mathbb{N}\}$, and hence for some k, F_k is uncountable. So, without loss of generality, we may suppose that $|\sigma| = |\tau|$ whenever $\sigma, \tau \in F$, say $|\sigma| = k$ $(\sigma \in F)$. The proof is by induction on k. The case $k = 1$ is trivial.

Suppose that the result holds for k and that $|\sigma| = k + 1$ $(\sigma \in F)$. Define $f : \sigma \mapsto \sup \sigma$, $F \to \omega_1$. We *claim* that $\{f(\sigma) : \sigma \in F\}$ is uncountable. For otherwise, there exists $\tau < \omega_1$ such that $\sigma \subset [0,\tau]$ $(\sigma \in F)$. But there are only countably many distinct finite subsets of $[0,\tau]$, and this contradicts the fact that F is uncountable.

Choose $F^* \subset F$ so that F^* is uncountable and $f|F^*$ is injective, and set $F^{**} = \{\sigma\backslash\{f(\sigma)\} : \sigma \in F^*\}$.

(i) Suppose that F^{**} is countable. Choose $G \subset F^*$ and a finite subset T of ω_1 such that G is uncountable and $\sigma\backslash\{f(\sigma)\} = \tau\backslash\{f(\tau)\} = T$.

(ii) Suppose that F^{**} is uncountable. We can suppose that $\sigma \mapsto \sigma\backslash\{f(\sigma)\}$, $F^* \to F^{**}$, is injective. For $\tau \in F^{**}$, $|\tau| = k$, and so by the inductive hypothesis, there exists $G^{**} \subset F^{**}$ and a finite subset T of ω_1 such that G^{**} is uncountable and $A \cap B = T$ whenever A and B are distinct elements of G^{**}. Set $G = \{\sigma \in F^* : \sigma\backslash\{f(\sigma)\} \in G^{**}\}$.

In either case, G and T are as required, and so the induction continues. ∎

6.2 COROLLARY

Let κ be a cardinal. Then $\{0,1\}^{\kappa}$ is ccc.

Proof

A basic open set U in $\{0,1\}^\kappa$ is specified by a finite subset F_U of κ and a map $\phi_U : F_U \to \{0,1\}$.

Let $\mathcal{F}$ be an uncountable family of basic open sets in $\{0,1\}^\kappa$. By 6.1, there is a finite subset T of κ and an uncountable subfamily $\mathcal{G}$ of $\mathcal{F}$ such that $F_U \cap F_V = T$ whenever $U,V \in \mathcal{G}$ with $U \neq V$. We may suppose that $\phi_U|T = \phi_V|T$, and so $U \cap V \neq \emptyset$ $(U,V \in \mathcal{G})$. Thus $\{0,1\}^\kappa$ is ccc. ∎

We can now turn to the main theory of this chapter.

6.2 DEFINITION

Let $(P,<)$ be a partially ordered set, and let A and B be subsets of P. Then A and B are <u>mutually cofinal</u> (respectively, <u>mutually coinitial</u>) if, for each $a \in A$, $b \in B$, there exist $c \in A$, $d \in B$ with $a \leqslant d$ and $b \leqslant c$ (respectively, $d \leqslant a$ and $c \leqslant b$). The set A is <u>cofinal</u> (respectively, <u>coinitial</u>) in P if A and P are mutually cofinal (respectively, mutually coinitial). The <u>cofinality</u> (respectively, <u>coinitiality</u>) of P is the minimum cardinal κ such that there is a cofinal (respectively, coinitial) subset of P with cardinality κ.

For convenience, we regard the empty set as having cofinality and coinitiality 0. Note that the cofinality of a subset of a totally ordered set is necessarily either 0, 1, or an infinite cardinal.

For example, let P be a non-empty subset of $\mathbb{R}$. Then the cofinality and coinitiality of P are each either 1 or $\aleph_0$.

Let A and B be subsets of a partially ordered set $(P,<)$. We write

$$A < B$$

if $a < b$ for each $a \in A$ and $b \in B$. An element c of P

interpolates the pair $\{A,B\}$ if

$$A < \{c\} < B.$$

6.4 DEFINITION

An ordered pair $\langle A,B\rangle$ of subsets of a partially ordered set $(P,<)$ is a pregap in P if $A \cup B$ is totally ordered and if $A < B$; the pregap $\langle A,B\rangle$ is a gap if no element P interpolates $\{A,B\}$. Two pregaps $\langle A_1,B_1\rangle$ and $\langle A_2,B_2\rangle$ are equivalent if A_1 and A_2 are mutually cofinal and B_1 and B_2 are mutually coinitial. A pregap $\langle A,B\rangle$ is a (κ,λ)-pregap if A has cofinality κ and B has coinitiality λ.

Suppose that $A = \{a_\alpha : \alpha < \kappa\}$ and that $B = \{b_\beta : \beta < \lambda\}$ are two sets such that

$$a_{\alpha_1} < a_{\alpha_2} < b_{\beta_2} < b_{\beta_1} \tag{1}$$

whenever $\alpha_1 < \alpha_2 < \kappa$ and $\beta_1 < \beta_2 < \lambda$. Then we shall abuse notation by writing

$$\langle a_\alpha, b_\beta : \alpha < \kappa,\ \beta < \lambda\rangle \tag{2}$$

for the pregap $\langle A,B\rangle$. If κ and λ are regular cardinals, then $\langle A,B\rangle$ is a (κ,λ)-pregap. When we write a pregap in the form (2), we implicitly assume that condition (1) is satisfied.

6.5 DEFINITION

A partially ordered set P is an η_1-set if, for each $\kappa,\lambda \leqslant \aleph_0$, there are no (κ,λ)-gaps.

Thus, P is an η_1-set if and only if, for each pair $\{A,B\}$ of countable subsets of P such that $A \cup B$ is totally ordered and $A < B$, there exists $c \in P$ with $A < \{c\} < B$. The sets A or B may be the empty set, and so, in an η_1-set, no countable chain is either cofinal or coinitial.

For example, $(\mathbb{R},<)$ is not an η_1-set: the pregap $\langle\{0\}, \{1/n : n \in \mathbb{N}\}\rangle$ is a $(1,\aleph_0)$-gap.

Let $(K,<)$ be an ordered field. Then $(K,<)$ is an <u>η_1-field</u> if it is an η_1-set. Let $\mathcal{U}$ be a free ultrafilter on $\mathbb{N}$. Then the ultraproduct $(\mathbb{R}^{\mathbb{N}}/\mathcal{U},<_{\mathcal{U}})$ is an η_1-field: this is an easy proof, similar to that given in 6.6, below. Thus we see a key difference between the two real-closed ordered fields $\mathbb{R}$ and $\mathbb{R}^{\mathbb{N}}/\mathcal{U}$. The subalgebras $c_o/\mathcal{U}$ and $(\ell^\infty/\mathcal{U})^*$ of $\mathbb{R}^{\mathbb{N}}/\mathcal{U}$ are also η_1-sets.

Recall from 2.2(vi) that $f \ll_F g$ in $\mathbb{N}^{\mathbb{N}}$ if $g(n) - f(n) \to \infty$ as $n \to \infty$, and from 3.1, that, for $g \in \mathbb{N}^{\mathbb{N}}$ with $1 \ll_F g$,

$$\langle g\rangle = \{f \in \mathbb{N}^{\mathbb{N}} : 1 \ll_F f \ll_F g\}.$$

6.6 PROPOSITION

For each $g \in \mathbb{N}^{\mathbb{N}}$ with $1 \ll_F g$, $(\langle g\rangle,\ll_F)$ is an η_1-set.

<u>Proof</u>

Let $\langle f_n,g_n : n \in \omega\rangle$ be an $(\aleph_o,\aleph_o)$-pregap in $(\langle g\rangle,\ll_F)$. Take $k_o = 1$, and, for each $n \in \mathbb{N}$, choose $k_n \in \mathbb{N}$ such that $k_n > k_{n-1}$ and

$$f_i(k) + 2n \leqslant g_j(k) \quad (i,j = 1,\dots,n,\ k \geqslant k_n).$$

Set $h(k) = 1 \quad (k = 1,\dots,k_1-1)$, and set

$$h(k) = \max\{f_1(k),\dots,f_n(k)\} + n \quad (k = k_n,\dots,k_{n+1}-1).$$

Then h interpolates $\langle f_n,g_n : n \in \omega\rangle$, and so there are no $(\aleph_o,\aleph_o)$-gaps in $(\langle g\rangle,\ll_F)$.

A similar argument covers the other cases of the claim that, for each $\kappa,\lambda \leqslant \aleph_o$, there are no (κ,λ)-gaps in $(\langle g\rangle,\ll_F)$. ∎

It is similarly easy to show that there are no $(\aleph_0,\aleph_0)$-gaps, no $(1,\aleph_0)$-gaps, and no $(\aleph_0,1)$-gaps in $(\mathbb{N}^{\mathbb{N}},<_F)$. However there are obvious (1,1)-gaps: it is to avoid this inconvenience that we sometimes work with the strong Fréchet order $<<_F$.

Let us now consider $(\aleph_1,\aleph_1)$-gaps in $(\mathbb{N}^{\mathbb{N}},<_F)$. It is easy to see in the theory ZFC + CH that there are $(\aleph_1,\aleph_1)$-gaps in $(\mathbb{N}^{\mathbb{N}},<_F)$: this holds essentially because $\mathbb{N}^{\mathbb{N}}$ can then be enumerated as a sequence of length ω_1. It is perhaps surprising that it is a theorem of ZFC that such $(\aleph_1,\aleph_1)$-gaps exist. This result is closely related to a famous theorem of Hausdorff that there are $(\aleph_1,\aleph_1)$-gaps in the partially ordered set $(\mathcal{P}(\mathbb{N}),<_F)$ which was introduced in Example 2.2(ii). We shall prove the result assuming MA + ¬CH.

In the next results, $\leqslant$ as a relation between functions denotes the standard order introduced in 2.2(iv).

6.7 DEFINITION

A <u>Hausdorff gap</u> is a pregap $<A,B>$ in $(\mathbb{N}^{\mathbb{N}},<_F)$ with the property that there exist $A_1 = \{f_\alpha : \alpha < \omega_1\}$ and $B_1 = \{g_\beta : \beta < \omega_1\}$ such that $<A_1,B_1>$ is an equivalent pregap to $<A,B>$, such that $f_{\alpha_1} <_F f_{\alpha_2} <_F g_{\beta_2} <_F g_{\beta_1}$ whenever $\alpha_1 < \alpha_2 < \omega_1$ and $\beta_1 < \beta_2 < \omega_1$, such that $f_\alpha \leqslant g_\alpha$ $(\alpha < \omega_1)$, and such that, if $\alpha < \beta < \omega_1$, then either $f_\alpha \not\leqslant g_\beta$ or $f_\beta \not\leqslant g_\alpha$.

Proposition 6.8, below, shows that the pregap $<A_1,B_1>$ is a gap, and so a Hausdorff gap is a special type of $(\aleph_1,\aleph_1)$-gap. These gaps may appear to be somewhat bizarre, but they will play their rôle when we establish the relative consistency of NDH in Chapter 8.

6.8 PROPOSITION

Each Hausdorff gap in $(\mathbb{N}^{\mathbb{N}},<_F)$ is a gap.

<u>Proof</u>

Let $\langle f_\alpha, g_\alpha : \alpha < \omega_1\rangle$ be as specified above, and, if possible, let h interpolate the pregap $\langle f_\alpha, g_\alpha\rangle$.

For each $\alpha < \omega_1$, there exists $n_\alpha \in \mathbb{N}$ such that $f_\alpha(k) < h(k) < g_\alpha(k)$ for $k \geq n_\alpha$. Necessarily, there exists $m_o \in \mathbb{N}$ such that $n_\alpha = m_o$ for uncountably many α. Since the set $\mathbb{N}^{m_o}$ is countable, there exists $x \in \mathbb{N}^{m_o}$ and an uncountable set S such that $f_\alpha | m_o = x$ and $n_\alpha = m_o$ $(\alpha \in S)$. Take $\alpha, \beta \in S$. For $k \leq m_o$, $f_\alpha(k) = f_\beta(k) \leq g_\beta(k)$, and, for $k > m_o$, $f_\alpha(k) < h(k) < g_\beta(k)$, and so $f_\alpha \leq g_\beta$. Similarly $f_\beta \leq g_\alpha$. But this contradicts the definition of a Hausdorff gap.

Thus no h interpolates the pregap. ∎

It may not be clear that Hausdorff gaps exist. We now prove that, with $MA + \neg CH$, there are Hausdorff, and hence $(\aleph_1, \aleph_1)$-gaps, in $(\mathbb{N}^{\mathbb{N}}, <_F)$.

Let $j, k \in \mathbb{N}$ with $j \leq k$, let $q_1 = (f_1, g_1) \in \mathbb{N}^j \times \mathbb{N}^j$, and let $q_2 = (f_2, g_2) \in \mathbb{N}^k \times \mathbb{N}^k$. Then q_2 <u>extends</u> q_1 if $f_2|j = f_1$ and $g_2|j = g_1$.

6.9 THEOREM $(MA + \neg CH)$

There is a Hausdorff gap in $(\mathbb{N}^{\mathbb{N}}, <_F)$.

<u>Proof</u>

Let P be the family of functions p whose domain, dom p, is a finite subset of ω_1 and whose range is a subset of $\mathbb{N}^k \times \mathbb{N}^k$ for some $k \in \mathbb{N}$, say $k = r(p)$. For $\alpha \in$ dom p, we write $p(\alpha) = (f_\alpha^p, g_\alpha^p)$.

Let Q be the set of elements q of P such that:

(i) if $\alpha \in$ dom q, then $f_\alpha^q \leq g_\alpha^q$ in $\mathbb{N}^{r(q)}$;

(ii) if $\alpha, \beta \in$ dom q with $\alpha \neq \beta$, then either $f_\alpha^q \not\leq g_\beta^q$ or $f_\beta^q \not\leq g_\alpha^q$ in $\mathbb{N}^{r(q)}$.

For $q_1, q_2 \in Q$, set $q_1 \leq q_2$ if:

(iii) dom $q_1 \supset$ dom q_2 and $r(q_1) \geq r(q_2)$;

(iv) if $\alpha \in \mathrm{dom}\, q_2$, then $q_1(\alpha)$ extends $q_2(\alpha)$;

(v) if $\alpha, \beta \in \mathrm{dom}\, q_2$ with $\alpha < \beta$, and if $q_1(\alpha) = (f_\alpha, g_\alpha)$ and $q_1(\beta) = (f_\beta, g_\beta)$, then

$$f_\alpha(j) < f_\beta(j) < g_\beta(j) < g_\alpha(j)$$

for $j = r(q_2)+1, \ldots, r(q_1)$.

It is easily checked that $(Q,<)$ is a partially ordered set.

We next make a *remark*. Take $q_1, q_2 \in Q$ with $r(q_1) = r(q_2)$, and let $T = (\mathrm{dom}\, q_1) \cap (\mathrm{dom}\, q_2)$. Suppose that $q_1|T = q_2|T$ and that $k \in \mathbb{N}$. Then there exists $q_3 \in Q$ with $q_3 \leqslant q_1$, $q_3 \leqslant q_2$, and $r(q_3) \geqslant k$. For we may obtain q_3 to satisfy these conditions by extending all the elements of q_1 and of q_2 appropriately, maintaining (v) and obtaining (ii).

We can now prove that $(Q,<)$ is ccc. For suppose, if possible, that C is an uncountable antichain in Q. We may suppose that there exists $m \in \mathbb{N}$ such that $r(q) = m \quad (q \in C)$. For each finite subset S of ω_1, there are only countably many elements q of Q with $\mathrm{dom}\, q = S$, and so we can suppose that $\mathrm{dom}\, q_1 \neq \mathrm{dom}\, q_2$ if q_1 and q_2 are distinct elements of C. By 6.1, there is a finite subset T of ω_1 and an uncountable subset C_1 of C such that $(\mathrm{dom}\, q_1) \cap (\mathrm{dom}\, q_2) = T$ whenever q_1 and q_2 are distinct elements of C_1. Again, there are only countably many choices of $q|T$, and so we may suppose that $q_1|T = q_2|T \quad (q_1, q_2 \in C_1)$. But now, by the remark, if $q_1, q_2 \in C_1$, then q_1 and q_2 are compatible, a contradiction. So $(Q,<)$ is ccc.

For $\alpha < \omega_1$ and $k \in \mathbb{N}$, set

$$D_{\alpha,k} = \{q \in Q : \alpha \in \mathrm{dom}\, q,\ r(q) \geqslant k\}.$$

It follows from the remark that each $D_{\alpha,k}$ is dense in $(Q,<)$. Clearly $|\{D_{\alpha,k} : \alpha < \omega_1,\ k \in \mathbb{N}\}| = \aleph_1 < 2^{\aleph_0}$, and hence, by MA, there is a filter G in Q with $G \cap D_{\alpha,k} \neq \emptyset$ for each $\alpha < \omega_1$ and each $k \in \mathbb{N}$.

Take $\alpha < \omega_1$. Since G is a filter, there are well-defined sequences $f_\alpha, g_\alpha \in \mathbb{N}^{\mathbb{N}}$ such that, for each $k \in \mathbb{N}$, $(f_\alpha|k, g_\alpha|k) = q(\alpha)$ for some $q \in G \cap D_{\alpha,k}$.

Take $\alpha < \beta < \omega_1$. It follows from (v) that $f_\alpha <_F f_\beta <_F g_\beta <_F g_\alpha$, and so $\langle f_\alpha, g_\alpha : \alpha < \omega_1 \rangle$ is a pregap in $(\mathbb{N}^{\mathbb{N}}, <_F)$. It follows from (i) and (ii) that this pregap is a Hausdorff gap. ∎

In fact, the partially ordered set Q of the above proof is sufficiently nice that it can be proved in ZFC that there is a filter G in Q with $G \cap D_{\alpha,k} \neq \emptyset$ in each case. Thus Theorem 6.9 is actually a result of ZFC.

It can also be shown that, with MA + ¬CH, the only (κ,λ)-gaps in $(\mathbb{N}^{\mathbb{N}}, <_F)$ with $\kappa, \lambda < 2^{\aleph_0}$ are $(1,1)$-gaps and $(\aleph_1, \aleph_1)$-gaps. This illustrates the special nature of $(\aleph_1, \aleph_1)$-gaps. The result can be proved by a variation of 6.15, below.

Let $f, g \in \mathbb{N}^{\mathbb{N}}$. Then $f =_F g$ if $f(n) = g(n)$ eventually. Set $(f_1, g_1) \sim (f_2, g_2)$ in $\mathbb{N}^{\mathbb{N}} \times \mathbb{N}^{\mathbb{N}}$ if $f_1 =_F f_2$ and $g_1 =_F g_2$. Clearly $\sim$ is an equivalence relation on $\mathbb{N}^{\mathbb{N}} \times \mathbb{N}^{\mathbb{N}}$; the equivalence class containing (f,g) is $[(f,g)]$. It is important to note that each such equivalence class is countable.

6.10 DEFINITION

Let P be the family of non-empty, finite subsets p of $\mathbb{N}^{\mathbb{N}} \times \mathbb{N}^{\mathbb{N}}$ such that:

(i) if $(f,g) \in p$, then $f \leqslant g$ in $\mathbb{N}^{\mathbb{N}}$;

(ii) if (f_1,g_1), (f_2,g_2) are distinct elements of p, then either $f_1 \not\leqslant g_2$ or $f_2 \not\leqslant g_1$.

For $p_1, p_2 \in P$, set $p_1 \leqslant p_2$ if $p_1 \supset p_2$.

Certainly $(P,<)$ is a partially ordered set.

Now let $\langle f_\alpha, g_\alpha : \alpha < \omega_1\rangle$ be an $(\aleph_1, \aleph_1)$-pregap in $(\mathbb{N}^{\mathbb{N}}, <_F)$, and set $E_\alpha = [(f_\alpha, g_\alpha)]$ for $\alpha < \omega_1$. Then $E_\alpha \cap E_\beta = \emptyset$ if $\alpha \neq \beta$.

6.11 DEFINITION

Let $\langle f_\alpha, g_\alpha : \alpha < \omega_1\rangle$ be an $(\aleph_1, \aleph_1)$-pregap in $(\mathbb{N}^{\mathbb{N}}, <_F)$. Then

$$Q(\langle f_\alpha, g_\alpha\rangle)$$

is the suborder of P consisting of those elements q of P such that $q \subset \bigcup\{E_\alpha : \alpha < \omega_1\}$, such that $|q \cap E_\alpha| \leq 1$ $(\alpha < \omega_1)$, and such that $\{\alpha : q \cap E_\alpha \neq \emptyset\} = \text{supp}\, q$ is a finite, nonempty set. For $\alpha \in \text{supp}\, q$, we denote by $q(\alpha)$ the unique element of $q \cap E_\alpha$.

Thus the set $Q(\langle f_\alpha, g_\alpha\rangle)$ contains those elements q of P formed by specifying a nonempty, finite subset supp q of ω_1 and by choosing an element $q(\alpha)$ of E_α for each $\alpha \in \text{supp}\, q$. The partially ordered set $Q(\langle f_\alpha, g_\alpha\rangle)$ will play a crucial rôle in Chapter 8: because of this we attempt an explanation of the intuition behind the definition. Suppose we could find a pregap $\langle f^*_\alpha, g^*_\alpha : \alpha < \omega_1\rangle$ such that $f^*_\alpha \leq g^*_\alpha$ $(\alpha < \omega_1)$, such that $f^*_\alpha \not\leq g^*_\beta$ $(\alpha, \beta < \omega_1,\ \alpha \neq \beta)$ and such that $(f^*_\alpha, g^*_\alpha) \sim (f_\alpha, g_\alpha)$ $(\alpha < \omega_1)$. Then $\langle f^*_\alpha, g^*_\alpha : \alpha < \omega_1\rangle$ is equivalent to $\langle f_\alpha, g_\alpha : \alpha < \omega_1\rangle$ and so $\langle f_\alpha, g_\alpha : \alpha < \omega_1\rangle$ is a Hausdorff gap. The elements q of $Q(\langle f_\alpha, g_\alpha\rangle)$ are "finite approximations" to the required pregap $\langle f^*_\alpha, g^*_\alpha\rangle$.

There are three easily verified remarks about $Q(\langle f_\alpha, g_\alpha\rangle)$. First, for each non-empty, finite subset F of ω_1, there is an element $q \in Q(\langle f_\alpha, g_\alpha\rangle)$ with supp q = F. Second, $|Q(\langle f_\alpha, g_\alpha\rangle)| = \aleph_1$. Third, $Q(\langle f_\alpha, g_\alpha\rangle)$ is a separative partially ordered set (cf. page 31).

A final fact about $Q(\langle f_\alpha, g_\alpha\rangle)$ is very important, but less easy. It applies in the case where $\langle f_\alpha, g_\alpha\rangle$ is actually a gap.

6.12 THEOREM

Let $\langle f_\alpha, g_\alpha : \alpha < \omega_1\rangle$ be an $(\aleph_1, \aleph_1)$-gap in $(\mathbb{N}^{\mathbb{N}}, <_F)$. Then $Q(\langle f_\alpha, g_\alpha\rangle)$ is a ccc partially ordered set.

<u>Proof</u>

Write Q for $Q(\langle f_\alpha, g_\alpha\rangle)$. Suppose, if possible, that C is an uncountable antichain in Q.

For each finite subset F of ω_1, there are only countably many elements q of Q with $\text{supp } q = F$ because each E_α is countable, and so we may suppose that $\text{supp } q_1 \neq \text{supp } q_2$ if q_1 and q_2 are distinct elements of C. By 6.1, there is a finite subset T of ω_1 and an uncountable subset C_1 of C such that $\text{supp } q_1 \cap \text{supp } q_2 = T$ whenever q_1 and q_2 are distinct elements of C_1. Again, there are only countably many choices of $q(T)$, and so we may suppose that $q_1(\alpha) = q_2(\alpha)$ $(q_1, q_2 \in C_1, \alpha \in T)$.

Let q_1 and q_2 be distinct elements of C_1, and let $q_3 = q_1 \cup q_2$. Since q_1 and q_2 are incompatible, $q_3 \notin Q$. Thus q_3 must fail to satisfy condition 6.10(ii), for it does satisfy all the other conditions for membership of Q. Hence there are distinct elements (f_1, g_1) and (f_2, g_2) of q_3 with $f_1 \leq g_2$ and $f_2 \leq g_1$. Necessarily, one of (f_1, g_1) and (f_2, g_2) belongs to $q_1 \backslash q_1(T)$, the other to $q_2 \backslash q_2(T)$. Then $q_1 \backslash q_1(T)$ and $q_2 \backslash q_2(T)$ are incompatible elements of Q, and so, replacing each $q \in C_1$ by $q \backslash q(T)$, we may suppose that $\text{supp } q_1 \cap \text{supp } q_2 = \emptyset$ whenever q_1 and q_2 are distinct elements of C_1.

For each $q \in C_1$ and $k \in \mathbb{N}$, set

$$F_q(k) = \min\{f(k) : (f,g) \in q \text{ for some } g \in \mathbb{N}^{\mathbb{N}}\},$$

$$G_q(k) = \max\{g(k) : (f,g) \in q \text{ for some } f \in \mathbb{N}^{\mathbb{N}}\}.$$

Clearly $F_q \leq G_q$ $(q \in C_1)$. Further, if q_1 and q_2 are distinct elements of C_1, then $q_1 \cup q_2$ fails to satisfy condition 6.10(ii), and so there exist $(f_1, g_1) \in q_1$ and $(f_2, g_2) \in q_2$ with $f_1 \leq g_2$. Thus $F_{q_1} \leq G_{q_2}$. For $k \in \mathbb{N}$, set

$$h(k) = \max\{F_q(k) : q \in C_1\}.$$

Then $h \in \mathbb{N}^{\mathbb{N}}$ and $F_q \leq h \leq G_q$ for each $q \in C_1$.

Since $\{\text{supp } q : q \in C_1\}$ is an uncountable family of pairwise disjoint, non-empty subsets of ω_1, it follows that, for each $\xi < \omega_1$, there exists $q_\xi \in C_1$ with $\text{supp } q_\xi \subset (\xi,\omega_1)$. Take $\xi < \omega_1$. Then there exists $\alpha \in \text{supp } q_\xi$ such that $F_{q_\xi}(n) = f_\alpha(n)$ eventually, and so $f_\xi <_F F_{q_\xi}$. Thus $f_\xi <_F h$. Similarly, $h <_F g_\xi$ $(\xi < \omega_1)$, and so h interpolates the gap $\langle f_\alpha,g_\alpha : \alpha < \omega_1\rangle$ in $(\mathbb{N}^{\mathbb{N}},<_F)$, a contradiction.

Thus Q is ccc, as required. ∎

6.13 THEOREM

Let $\langle f_\alpha,g_\alpha : \alpha < \omega_1\rangle$ be an $(\aleph_1,\aleph_1)$-pregap in $(\mathbb{N}^{\mathbb{N}},<_F)$, and set $Q = Q(\langle f_\alpha,g_\alpha\rangle)$. For each $\xi < \omega_1$, set $D_\xi = \{q \in Q : \xi \in \text{supp } q\}$, and set $\mathcal{D} = \{D_\xi : \xi < \omega_1\}$. Then each D_ξ is dense in Q. If there is a $\mathcal{D}$-generic filter on Q, then $\langle f_\alpha,g_\alpha : \alpha < \omega_1\rangle$ is a Hausdorff gap.

Proof

Take $\xi < \omega_1$, and take $q \in Q\backslash D_\xi$. Then there exists $(f,g) \in E_\xi$ with $f \leq g$ and such that $f \not\leq k$ for each $(h,k) \in q$. Define $\bar{q} \in Q$ such that $\bar{q}(\alpha) = q(\alpha)$ $(\alpha \in \text{supp } q)$ and $\bar{q}(\xi) = (f,g)$. Then $\bar{q} \in D_\xi$ and $\bar{q} \leq q$ in Q. Thus D_ξ is dense in $(Q,<)$.

Now suppose that G is a $\mathcal{D}$-generic filter on Q, so that $G \cap D_\xi \neq \emptyset$ $(\xi < \omega_1)$.

Take $q_1,q_2 \in G \cap D_\xi$. Since G is a filter, there exists $q \in Q$ with $q \supset q_1 \cup q_2$, and so $q_1(\xi) = q_2(\xi)$. Thus $q(\xi)$ is the same - say (f^*_ξ,g^*_ξ) - for each $q \in G \cap D_\xi$.

Certainly $f^*_\xi \leq g^*_\xi$ $(\xi < \omega_1)$. Take $\xi,\eta < \omega_1$ with $\xi \neq \eta$, take $q_1 \in G \cap D_\xi$, $q_2 \in G \cap D_\eta$, and take $q \supset q_1 \cup q_2$. Since q satisfies 6.10(ii), either $f^*_\xi \not\leq g^*_\eta$ or $f^*_\eta \not\leq g^*_\xi$. Thus $\langle f^*_\xi,g^*_\xi : \xi < \omega_1\rangle$ is a Hausdorff gap in

$(\mathbb{N}^{\mathbb{N}},<_F)$. Since $(f^*_\xi,g^*_\xi) \sim (f_\xi,g_\xi)$ $(\xi < \omega_1)$, the gap $\langle f^*_\xi,g^*_\xi : \xi < \omega_1\rangle$ is equivalent to $\langle f_\xi,g_\xi : \xi < \omega_1\rangle$.

This completes the proof of the theorem. ∎

6.14 COROLLARY (MA + ¬CH)

Each $(\aleph_1,\aleph_1)$-gap in $(\mathbb{N}^{\mathbb{N}},<_F)$ is a Hausdorff gap.

Proof

Let $\langle f_\alpha,g_\alpha : \alpha < \omega_1\rangle$ be an $(\aleph_1,\aleph_1)$-gap, and set $Q = Q(\langle f_\alpha,g_\alpha\rangle)$. By 6.12, Q is ccc. Let $\mathcal{D}$ be as in 6.13. Then $|\mathcal{D}| = \aleph_1$, and so $|\mathcal{D}| < 2^{\aleph_0}$. By MA, there is a $\mathcal{D}$-generic filter on Q, and so the result follows from 6.13. ∎

We now come to a key theorem. The result is perhaps surprising: it shows that, with MA, each (κ,λ)-pregap in $(\mathbb{N}^{\mathbb{N}},<_F)$ with $\kappa,\lambda < 2^{\aleph_0}$ can be "interpolated by two functions". Since there are $(\aleph_1,\aleph_1)$-gaps in $(\mathbb{N}^{\mathbb{N}},<_F)$, one function is not always sufficient.

6.15 THEOREM (MA)

Let $\langle f_\alpha,g_\beta : \alpha < \kappa,\ \beta < \lambda\rangle$ be a (κ,λ)-pregap in $(\mathbb{N}^{\mathbb{N}},<_F)$ with $\kappa,\lambda < 2^{\aleph_0}$. Then there exist two functions h_1 and h_2 in $\mathbb{N}^{\mathbb{N}}$ such that, for each $\alpha < \kappa$ and $\beta < \lambda$, there exists $k \in \mathbb{N}$ such that, for each $j \geq k$, either $f_\alpha(j) \leq h_1(j) \leq g_\beta(j)$ or $f_\alpha(j) \leq h_2(j) \leq g_\beta(j)$.

Proof

Let P be the set of triples $p = \langle r,s,\sigma\rangle$ such that:

(i) $r,s \in \mathbb{N}^{\ell(p)}$ for some $\ell(p) \in \mathbb{N}$;

(ii) σ is a finite subset of $\kappa \times \lambda$;

(iii) for each $(\alpha,\beta) \in \sigma$ and each $j > \ell(p)$, $f_\alpha(j) < g_\beta(j)$.

For $p_1 = \langle r_1, s_1, \sigma_1\rangle$ and $p_2 = \langle r_2, s_2, \sigma_2\rangle$ in P, set $p_1 \leqslant p_2$ if:

(iv) $\sigma_1 \supset \sigma_2$ and $\ell(p_1) \geqslant \ell(p_2)$;

(v) (r_1, s_1) extends (r_2, s_2);

(vi) for each $(\alpha,\beta) \in \sigma_2$ and each $j \in \{\ell(p_2)+1,\ldots,\ell(p_1)\}$, either $r_1(j) \in [f_\alpha(j), g_\beta(j)]$ or $s_1(j) \in [f_\alpha(j), g_\beta(j)]$.

It is easily checked that $(P,<)$ is a partially ordered set.

Let $p_1 = \langle r, s, \sigma_1\rangle$, $p_2 = \langle r, s, \sigma_2\rangle \in P$ (with $\ell(p_1) = \ell(p_2)$), and let $m \in \mathbb{N}$. We *claim* that there exists $p_3 \in P$ with $p_3 \leqslant p_1$, $p_3 \leqslant p_2$, and $\ell(p_3) \geqslant m$. For let $\sigma_3 = \sigma_1 \cup \sigma_2$, and choose $k \geqslant \max\{\ell(p_1), m\}$ such that, for each $(\alpha,\beta) \in \sigma_3$ and each $j > k$, $f_\alpha(j) < g_\beta(j)$: this is possible because $f_\alpha <_F g_\beta$ for each α, β. Choose $r^* \in \mathbb{N}^k$ so that r^* extends r and so that $r^*(j) \in [f_\alpha(j), g_\beta(j)]$ for $j = \ell(p_1)+1,\ldots,k$ and for each $(\alpha,\beta) \in \sigma_1$: this is possible because p_1 satisfies (iii). Similarly, choose $s^* \in \mathbb{N}^k$ to extend s. Let $p_3 = (r^*, s^*, \sigma_3)$. Then p_3 has the required properties, and so the claim is established.

In particular, elements $\langle r, s, \sigma_1\rangle$ and $\langle r, s, \sigma_2\rangle$ in P are compatible. Since there are only countably many choices of (r,s) such that $\langle r, s, \sigma\rangle \in P$ for some σ, P is ccc.

For $\alpha < \kappa$, $\beta < \lambda$, and $k \in \mathbb{N}$, set

$$D_{\alpha,\beta,k} = \{p = \langle r, s, \sigma\rangle \in P : (\alpha,\beta) \in \sigma, \ell(p) \geqslant k\}.$$

It follows easily from the above claim that each $D_{\alpha,\beta,k}$ is dense in $(P,<)$. Since $\kappa, \lambda < 2^{\aleph_0}$, $|\{D_{\alpha,\beta,k}\}| < 2^{\aleph_0}$, and hence, by MA, there is a filter G in P with $G \cap D_{\alpha,\beta,k} \neq \emptyset$ in each case.

For $k \in \mathbb{N}$, there exists $q = (r, s, \sigma) \in G$ with $\ell(q) \geqslant k$. Set $h_1(k) = r(k)$, $h_2(k) = s(k)$. Since G is a filter in P, it follows from (v) that h_1 and h_2 are well-defined elements of $\mathbb{N}^{\mathbb{N}}$. Take $\alpha < \kappa$ and $\beta < \lambda$, and take $q \in D_{\alpha,\beta,1} \cap G$. For each $j \geq \ell(q) + 1$, there exists $\bar{q} \in G$ with $\bar{q} \leq q$ and $\ell(\bar{q}) \geq j$, and so it follows from (vi) that either $h_1(j) \in [f_\alpha(j), g_\beta(j)]$ or $h_2(j) \in [f_\alpha(j), g_\beta(j)]$. This proves that $\{h_1, h_2\}$ has the required property. ∎

6.16 THEOREM (MA)

Let $\langle A,B \rangle$ be a (κ,λ)-pregap in $(\mathbb{N}^{\mathbb{N}}, <_F)$ with $\aleph_0 \leq \kappa, \lambda < 2^{\aleph_0}$, and let U be a free ultrafilter on $\mathbb{N}$. Then there exists h in $\mathbb{N}^{\mathbb{N}}$ such that $A <_U \{h\} <_U B$.

Proof

Suppose that $\langle A,B \rangle = \langle f_\alpha, g_\beta : \alpha < \kappa, \beta < \lambda \rangle$, and let h_1 and h_2 be the corresponding functions specified in 6.15. We *claim* that either $f_\alpha <_U h_1 <_U g_\beta$ $(\alpha < \kappa, \beta < \lambda)$ or that $f_\alpha <_U h_2 <_U g$ $(\alpha < \kappa, \beta < \lambda)$.

For suppose, if possible, that each of these two possibilities fails. Then there exist (α_1, β_1) and (α_2, β_2) in $\kappa \times \lambda$ such that the following two statements are both false: (i) $f_{\alpha_1} <_U h_1 <_U g_{\beta_1}$; (ii) $f_{\alpha_2} <_U h_2 <_U g_{\beta_2}$.

Take $\gamma = \max\{\alpha_1, \alpha_2\}$, and $\delta = \max\{\beta_1, \beta_2\}$. Then the statements $f_\gamma <_U h_1 <_U g_\delta$ and $f_\gamma <_U h_2 <_U g_\delta$ both fail.

Since $\kappa, \lambda \geq \aleph_0$, $\gamma + 1 < \kappa$ and $\delta + 1 < \lambda$. Set

$$A = \{j \in \mathbb{N} : f_{\gamma+1}(j) \leq h_1(j) \leq g_{\delta+1}(j)\},$$

$$B = \{j \in \mathbb{N} : f_{\gamma+1}(j) \leq h_2(j) \leq g_{\delta+1}(j)\}.$$

By Theorem 6.15, $\mathbb{N} \setminus (A \cup B)$ is finite, and so $A \cup B \in U$. Hence A or B, say A, belongs to U. Since $f_\gamma <_F f_{\gamma+1}$ and $g_{\delta+1} <_F g_\delta$, the set

$$\{j \in \mathbb{N} : f_\gamma(j) < h_1(j) < g_\delta(j)\}$$

belongs to $\mathcal{U}$, and so $f_\gamma <_{\mathcal{U}} h_1 <_{\mathcal{U}} g_\delta$, a contradiction.

Thus the claim holds, and the theorem follows. ∎

6.17 DEFINITION

Let $(P,<)$ be a partially ordered set. Then P is an $\underline{\alpha_1\text{-set}}$ if each non-empty subset of P has a countable cofinal and coinitial subset, and P is a $\underline{\beta_1\text{-set}}$ if there is an ordinal σ such that $P = \bigcup\{P_\nu : \nu < \sigma\}$, where $\{P_\nu : \nu < \sigma\}$ is a chain of α_1-sets in $(\mathcal{P}(P),\subset)$.

For example, it is easily checked that $(\mathbb{R},<)$ is an α_1-set.

Clearly, each subset of an α_1-set is an α_1-set, and the countable union of α_1-sets is an α_1-set. Each partially ordered set of cardinality $\aleph_1$ is a β_1-set.

6.18 PROPOSITION

Let $(P,<)$ be a totally ordered α_1-set, and let $(Q,<)$ be an η_1-set. Let P_o be a subset of P, and let $\pi_o : P_o \to Q$ be an isotonic map. Then there is an isotonic map $\pi : P \to Q$ such that $\pi|P_o = \pi_o$.

Proof

Let F be the family of pairs (U,ϕ) such that $P_o \subset U \subset P$ and $\phi : U \to Q$ is an isotonic map with $\phi|P_o = \pi_o$. Set $(U_1,\phi_1) \prec (U_2,\phi_2)$ in F if $U_1 \subset U_2$ and $\phi_2|U_1 = \phi_1$. Clearly, $(P_o,\pi_o) \in F$, and each chain in $(F,\prec)$ has an upper bound. By Zorn's Lemma, $(F,\prec)$ has a maximal element, say (V,π).

Suppose, if possible, that $V \neq P$, and take $a \in P\backslash V$. Since P is an α_1-set, there exist countable subsets P_1 and P_2 of P which are cofinal in the set $\{x \in P : x < a\}$ and coinitial in the set $\{x \in P : a < x\}$, respectively. Then $\pi(P_1)$ and $\pi(P_2)$ are countable sets with $\pi(P_1) \cup \pi(P_2)$ totally ordered and $\pi(P_1) < \pi(P_2)$ in Q. Since Q is an η_1-set, there exists $b \in Q$ with $\pi(P_1) < \{b\} < \pi(P_2)$. Set $\pi(a) = b$. Then $(V\cup\{a\},\pi) \succ (V,\pi)$

in F, a contradiction of the supposed maximality of (V,π). Thus $V = P$, and π is the required map. ∎

6.19 COROLLARY

Let $(P,<)$ be a totally ordered β_1-set, and let $(Q,<)$ be an η_1-set. Then there is an isotonic map from P into Q.

Proof

Let $P = \bigcup\{P_\nu : \nu < \sigma\}$, where $\{P_\nu : \nu < \sigma\}$ is a chain of α_1-sets in $(P(P),\subset)$. It follows from the proposition by induction that there is a family $\{\pi_\nu : \nu < \sigma\}$ of isotonic maps $P_\nu \to Q$ such that $\pi_\nu|P_\mu = \pi_\mu$ $(\mu < \nu < \sigma)$. Define $\pi : P \to Q$ by the formulae $\pi|P_\nu = \pi_\nu$ $(\nu < \sigma)$. Then π is well defined, and π is the required isotonic map. ∎

6.20 DEFINITION

The set $\mathbb{R}$ is the collection of sequences $x = \langle x_\alpha : \alpha < \omega_1\rangle$, where each $x_\alpha \in \{0,1\}$ and where $\{\alpha : x_\alpha = 0\}$ is a non-empty, proper subset of ω_1 which has no maximum element.

For distinct elements $x = \langle x_\alpha\rangle$ and $y = \langle y_\alpha\rangle$ in $\mathbb{R}$, set

$$\delta(x,y) = \inf\{\alpha : x_\alpha \neq y_\alpha\},$$

and set $x < y$ if $x(\delta(x,y)) < y(\delta(x,y))$.

Thus $\mathbb{R}$ is a subset of the totally ordered set of all binary sequences of length ω_1, and $<$ is the lexicographic order on $\mathbb{R}$. For example, let $x_\alpha = y_\alpha = 0$ $(\alpha < \omega)$, $x_\alpha = y_\alpha = 1$ $(\alpha > \omega)$, $x_\omega = 1$, $y_\omega = 0$. Then $\langle x_\alpha : \alpha < \omega_1\rangle \in \mathbb{R}$, but $\langle y_\alpha : \alpha < \omega_1\rangle \notin \mathbb{R}$.

6.21 DEFINITION

Take $\sigma \leq \omega_1$. The set $\mathbb{Q}_\sigma$ consists of those sequences $\langle x_\alpha : \alpha < \omega_1 \rangle$ such that there exists $\delta < \sigma$ with $x_\delta = 1$ and $x_\alpha = 0$ for $\alpha \in (\delta, \omega_1)$.

We write $\mathbb{Q}$ for $\mathbb{Q}_{\omega_1}$. Clearly

$$\mathbb{Q} = \bigcup\{\mathbb{Q}_\sigma : \sigma < \omega_1\},$$

and $\mathbb{Q} \subset \mathbb{R}$.

The properties of the ordered sets $(\mathbb{Q},<)$ and $(\mathbb{R},<)$ that are given in the next proposition will be used in Chapter 8. First, we give some further notation. Set $\mathbb{I} = \mathbb{R}\setminus\mathbb{Q}$, and set

$$\mathbb{J} = \{x \in \mathbb{R} : x \text{ is not eventually constant}\}.$$

Then $\mathbb{J} \subset \mathbb{I} \subset \mathbb{R}$. For $x \in \mathbb{I}$, set

$$A_x = \{y \in \mathbb{Q} : y < x\}, \quad B_x = \{y \in \mathbb{Q} : x < y\}. \tag{3}$$

Then $\langle A_x, B_x \rangle$ is a pregap in $(\mathbb{Q},<)$.

Let $(P,<)$ be a totally ordered set. A subset Q of P is <u>order-dense</u> in $(P,<)$ if, for each $a,b \in P$ with $a < b$, there exists $c \in Q$ with $a \leqslant c \leqslant b$.

6.22 PROPOSITION

(i) $|\mathbb{R}| = |\mathbb{J}| = 2^{\aleph_1}$, and $|\mathbb{Q}| = 2^{\aleph_0}$.

(ii) $(\mathbb{R},<)$ is complete, and each $(\aleph_1,\aleph_1)$-pregap in $(\mathbb{R},<)$ is interpolated.

(iii) $(\mathbb{Q},<)$ is order-dense in $(\mathbb{R},<)$.

(iv) If $x \in \mathbb{J}$, then $\langle A_x, B_x \rangle$ is an $(\aleph_1,\aleph_1)$-gap in $(\mathbb{Q},<)$, and, if $x \in \mathbb{I}\setminus\mathbb{J}$, then $\langle A_x, B_x \rangle$ is an $(\aleph_0,\aleph_1)$-gap or an $(\aleph_1,\aleph_0)$-gap in $(\mathbb{Q},<)$.

<u>Proof</u>

(i) Certainly $|\mathbb{R}| = |\mathbb{J}| = 2^{\aleph_1}$. For each

$\sigma < \omega_1$, $|\mathbb{Q}_\sigma| = 2^{\aleph_0}$, and so $2^{\aleph_0} \leqslant |\mathbb{Q}| \leqslant \aleph_1 . 2^{\aleph_0} = 2^{\aleph_0}$.

(ii) Let S be a non-empty subset of $\mathbb{R}$ which is bounded above. Define $x = \langle x_\alpha \rangle$ inductively by

$$x_\alpha = \max\{y_\alpha : y \in S,\ y|\alpha = x|\alpha\}.$$

If $x \in \mathbb{R}$, set $z = x$. If $x \notin \mathbb{R}$, then there exists $\sigma < \omega_1$ such that $x_\sigma = 0$ and $x_\xi = 1$ $(\xi \in (\sigma,\omega_1))$: set $z_\xi = x_\xi$ $(\xi < \sigma)$, $z_\sigma = 1$, $z_\xi = 0$ $(\xi \in (\sigma,\omega_1))$, and set $z = \langle z_\xi \rangle$. In either case $z = \sup S$. Similarly, each non-empty set in $\mathbb{R}$ which is bounded below has an infimum, and so $(\mathbb{R},<)$ is complete.

Let $\langle A,B \rangle$ be an $(\aleph_1,\aleph_1)$-pregap in $\mathbb{R}$, and set $h = \sup A$. Then h interpolates $\langle A,B \rangle$.

(iii) This is clear.

(iv) Suppose that $x = \langle x_\alpha : \alpha < \omega_1 \rangle \in \mathcal{I}$. Certainly $\langle A_x, B_x \rangle$ is a gap in $\mathbb{Q}$, say it is a (κ,λ)-gap.

Set $S = \{\xi : x_\xi = 1\}$. Then S is non-empty and has no maximum element. For $\xi \in S$, set $x_{\alpha\xi} = x_\alpha$ $(\alpha \leqslant \xi)$, $x_{\alpha\xi} = 0$ $(\alpha \in (\xi,\omega_1))$, and set $x^\xi = \langle x_{\alpha\xi} : \alpha < \omega_1 \rangle$. Then $\{x^\xi : \xi \in S\}$ is cofinal in A_x, and $\kappa = |S|$.

Set $T = \{\xi : x_\xi = 0\}$. Then T is non-empty and has no maximum element. For $\xi \in T$, set $y_{\alpha\xi} = x_\alpha$ $(\alpha < \xi)$, $y_{\xi\xi} = 1$, $y_{\alpha\xi} = 0$ $(\alpha \in (\xi,\omega_1))$, and set $y^\xi = \langle y_{\alpha\xi} : \alpha < \omega_1 \rangle$. Then $\{y^\xi : \xi \in T\}$ is coinitial in B_x, and $\lambda = |T|$.

If $x \in \mathcal{J}$, then $\sup S = \sup T = \omega_1$, and so $\kappa = \lambda = \aleph_1$. If $x \in \mathcal{I}\setminus\mathcal{J}$, then $\{\kappa,\lambda\} = \{\aleph_0,\aleph_1\}$. ∎

6.23 PROPOSITION

For each $\sigma < \omega_1$, $(\mathbb{Q}_\sigma,<)$ is an α_1-set, and $(\mathbb{Q},<)$ is a β_1-set.

Proof

Let A be a non-empty subset of $\mathbb{Q}_\sigma$. Clearly A is bounded above in $\mathbb{R}$. Let $x = \langle x_\alpha \rangle$ be the supremum of A in $\mathbb{R}$. If $x \in A$, then $\{x\}$ is cofinal in A. If $x \notin A$, set $\tau = \sup\{\alpha : x_\alpha = 1\}$. Then τ is a limit ordinal and $\tau \leqslant \sigma$. For each $\xi < \tau$, there exists $y^\xi \in A$ with $y^\xi|\xi = x|\xi$, and $\{y^\xi : \xi < \tau\}$ is a countable, cofinal subset in A.

Similarly, A has a countable, coinitial subset, and so $(\mathbb{Q}_\sigma,<)$ is an α_1-set.

Since $\mathbb{Q} = \bigcup\{\mathbb{Q}_\sigma : \sigma < \omega_1\}$, $\mathbb{Q}$ is a β_1-set. ∎

Recall that $Z(n) = n \ (n \in \mathbb{N})$, so that $1 \ll_F Z$.

6.24 THEOREM (MA + ¬CH)

Let U be a free ultrafilter on $\mathbb{N}$. Then there is an embedding of $(\mathbb{R},<)$ into $(\langle Z\rangle/U,<_U)$.

Proof

By 6.6, $(\langle Z\rangle,\ll_F)$ is an η_1-set, and, by 6.23, $(\mathbb{Q},<)$ is a β_1-set. Thus, by 6.19, there is an isotonic map $\pi : (\mathbb{Q},<) \to (\langle Z\rangle,\ll_F)$. The map $\pi : (\mathbb{Q},<) \to (\langle Z\rangle,<_F)$ is isotonic.

For each $x \in \mathbb{I}$, let $\langle A_x,B_x\rangle$ be the gap in $(\mathbb{Q},<)$ specified in (3). By 6.22(iv), $\langle A_x,B_x\rangle$ is a (κ,λ)-gap in $(\mathbb{Q},<)$ with $\{\kappa,\lambda\} \subset \{\aleph_0,\aleph_1\}$ and so $\langle\pi(A_x),\pi(B_x)\rangle$ is a (κ,λ)-pregap in $(\langle Z\rangle,<_F)$. Since $\aleph_1 < 2^{\aleph_0}$, it follows from 6.16 that there exists h in $\mathbb{N}^\mathbb{N}$ with $\pi(A_x) <_U \{h\} <_U \pi(B_x)$. Since $\pi(B_x) \neq \emptyset$, $h \in \langle Z\rangle$. Set $\pi(x) = h$. Then $\pi : (\mathbb{R},<) \to (\langle Z\rangle,<_U)$ is isotonic because $\mathbb{Q}$ is dense in $\mathbb{R}$, and π induces the required embedding $(\mathbb{R},<) \to (\langle Z\rangle/U,<_U)$. ∎

The main result of this chapter now follows immediately from Theorem 3.3.

6.25 THEOREM (MA + ¬CH)

Let X be a compact space. Assume that there is a discontinuous homomorphism from $C(X,\mathbb{C})$ into a Banach algebra. Then there is an embedding of $(\mathbb{R},<)$ into $(\mathbb{N}^\mathbb{N},<_F)$. ∎

To build a model of ZFC + NDH, it thus suffices to build a model of ZFC + MA + ¬CH in which there is no embedding of $(\mathbb{R},<)$ into $(\mathbb{N}^{\mathbb{N}},<_F)$. This we shall do in Chapter 8.

We conclude this chapter with some remarks (not required later) on the structure of some of the ordered sets that we have mentioned.

A partially ordered set $(P,<)$ is <u>ω-complete</u> if each countable set which is bounded above (respectively, below) has a supremum (respectively, an infimum). Set $\mathbb{Q}^{(\omega)} = \mathbb{R}\backslash J$, so that

$$\mathbb{Q}^{(\omega)} = \{x \in \mathbb{R} : x \text{ is eventually constant}\}.$$

Then $(\mathbb{Q}^{(\omega)},<)$ is easily checked to be the smallest ω-complete set containing $\mathbb{Q}$ as an order-dense subset, and $|\mathbb{Q}^{(\omega)}| = 2^{\aleph_0}$.

It is clear that $\mathbb{R}$ is the natural analogue of $\mathbb{R} \cap (0,1)$ "at the next cardinal", but it is arguable whether the natural analogue of $\mathbb{Q} \cap (0,1)$ is $\mathbb{Q}$ or $\mathbb{Q}^{(\omega)}$.

Certainly the sets $\mathbb{Q}$ and $\mathbb{R}$ are not order-isomorphic because $|\mathbb{Q}| < |\mathbb{R}|$. But, by 5.28(i), it could be that $|\mathbb{Q}| = |\mathbb{Q}^{(\omega)}| = |\mathbb{R}|$. In what way, then, do these three sets differ?

6.26 PROPOSITION

(i) $(\mathbb{Q},<)$ is both a β_1-set and an η_1-set.

(ii) $(\mathbb{Q}^{(\omega)},<)$ is a β_1-set, but not an η_1-set.

(iii) $(\mathbb{R},<)$ is neither a β_1-set nor an η_1-set.

<u>Proof</u>

(i) By 6.23, $\mathbb{Q}$ is a β_1-set.

Let A and B be countable sets in $\mathbb{Q}$ with $A < B$. Then there exists $\sigma < \omega_1$ with $A \cup B \subset \mathbb{Q}_\sigma$.

If $A = \emptyset$, take $x_\alpha = 0$ $(\alpha \neq \sigma)$, $x_\sigma = 1$, and set $x = \langle x_\alpha : \alpha < \omega_1 \rangle$. Then $\{x\} < B$. If $B = \emptyset$, take $x_\alpha = 1$ $(\alpha \leqslant \sigma)$, $x_\alpha = 0$ $(\alpha > \sigma)$, and set $x = \langle x_\alpha : \alpha < \omega_1 \rangle$. Then $A < \{x\}$. Thus we can suppose that both A and B are non-empty.

Let $x = \sup A$ and $y = \inf B$ in $(\mathbb{R}, <)$, so that $x < y$. Suppose, if possible, that there exists $\tau \geqslant \sigma$ with $x_\tau = 1$. Set $\bar{x}_\alpha = x_\alpha$ $(\alpha < \tau)$, $\bar{x}_\alpha = 0$ $(\alpha \geqslant \tau)$, and set $\bar{x} = \langle \bar{x}_\alpha : \alpha < \omega_1 \rangle$. Then $\bar{x} \in \mathbb{R}$, $\bar{x} < x$, and $A < \{\bar{x}\}$, a contradiction. So $x_\tau = 0$ $(\tau \geqslant \sigma)$. Similarly, $y_\tau = 1$ $(\tau \geqslant \sigma)$. Set $z_\alpha = x_\alpha$ $(\alpha < \sigma)$, $z_\sigma = 1$, $z_\sigma = 0$ $(\alpha > \sigma)$, and set $z = \langle z_\alpha : \alpha < \omega_1 \rangle$. Then $x < z < y$, and so $A < \{z\} < B$. Thus $\mathbb{Q}$ is an η_1-set.

(ii) The set $\mathbb{Q}^{(\omega)}$ is ω-complete, and so it is not an η_1-set. For $\sigma < \omega_1$, set

$$\mathbb{Q}_\sigma^{(\omega)} = \{\langle x_\alpha \rangle \in \mathbb{R} : x_\alpha = x_\sigma \quad (\alpha > \sigma)\}.$$

Then $\mathbb{Q}_{\sigma+1}$ is order-dense in $\mathbb{Q}_\sigma^{(\omega)}$, and so $\mathbb{Q}_\sigma^{(\omega)}$ is an α_1-set. Hence $\mathbb{Q}^{(\omega)} = \bigcup\{\mathbb{Q}_\sigma^{(\omega)} : \sigma < \omega_1\}$ is a β_1-set.

(iii) Again, $\mathbb{R}$ is clearly not an η_1-set; we prove that $\mathbb{R}$ is not a β_1-set.

For $x \in \{0,1\}^\tau$ and $\sigma \leqslant \tau$, set

$$U_{x,\sigma} = \{y \in \mathbb{R} : y|\sigma = x|\sigma\}.$$

Let S be an α_1-set in $\mathbb{R}$, and let x be a binary sequence of length σ, where $\sigma < \omega_1$. We *claim* that there exists $\tau \in (\sigma, \omega_1)$ and a binary sequence y of length τ extending x such that $y(\sigma) = 0$ and $U_{y,\tau} \cap S = \emptyset$. For $\zeta > \sigma$, set

$$y_\alpha^\zeta = x_\alpha \quad (\alpha < \sigma), \qquad y_\zeta^\zeta = 1,$$

$$y_\alpha^\zeta = 0 \quad (\sigma < \alpha < \zeta \text{ and } \zeta < \sigma < \omega_1),$$

and set $y^\zeta = \langle y_\alpha^\zeta : \alpha < \omega_1 \rangle$. Since S is an α_1-set, there exists ζ such that $[y^{\zeta+1},y^{\zeta}) \cap S = \emptyset$, for otherwise S would contain a strictly decreasing sequence of length ω_1. Take $\tau = \zeta + 2$, and set $y_\alpha = y_\alpha^{\zeta+1}$ $(\alpha < \tau)$. Since $U_{y,\tau} \subset [y^{\zeta+1},y^{\zeta})$, the claim holds.

To obtain a contradiction, suppose that $(\mathbb{R},<)$ is a β_1-set. Since $(\mathbb{Q},<)$ is an η_1-set, it follows from 6.19, that there is an isotonic map $\iota: \mathbb{R} \to \mathbb{Q}$. Set $\mathbb{R}_\zeta = \iota^{-1}(\mathbb{Q}_\zeta)$ $(\zeta < \omega_1)$. Then each $\mathbb{R}_\zeta$ is an α_1-set.

We construct recursively for $\zeta \leqslant \omega_1$ an ordinal $\sigma(\zeta) \leqslant \omega_1$ and a binary sequence x^ζ of length $\sigma(\zeta)$ as follows. Take $\sigma(0) = x(0) = 1$. Given $\sigma(\zeta)$ and x^ζ, take $\sigma(\zeta + 1) \in (\sigma(\zeta),\omega_1)$ and $x^{\zeta+1}$ of length $\sigma(\zeta + 1)$ extending x^ζ such that $x^{\zeta+1}(\sigma(\zeta)) = 0$ and $U_{x^{\zeta+1},\sigma(\zeta+1)} \cap \mathbb{R}_\zeta = \emptyset$: this is possible by the claim. For a limit ordinal ζ, take $\sigma(\zeta) = \sup\{\sigma(\xi) : \xi < \zeta\}$ and, for $\alpha < \sigma(\zeta)$, set $x_\alpha = x_\alpha^\xi$ whenever $\alpha < \sigma(\zeta)$.

Let $x = x^{\omega_1}$. Then $x \in \mathbb{R}$, but, for each $\zeta < \omega_1$, $x \notin \mathbb{R}_\zeta$. This is the required contradiction. ∎

6.27 COROLLARY

No two of the sets $(\mathbb{Q},<)$, $(\mathbb{Q}^{(\omega)},<)$, and $(\mathbb{R},<)$ are order-isomorphic. ∎

For each ultrafilter $\mathcal{U}$ on $\mathbb{N}$, $|\ell^\infty/\mathcal{U}| = 2^{\aleph_0}$, and so, with CH, $\ell^\infty/\mathcal{U}$ is a β_1-set. However, the following corollary shows that it cannot be proved in ZFC + ¬CH that there exists an ultrafilter $\mathcal{U}$ such that $\ell^\infty/\mathcal{U}$ is a β_1-set.

6.28 COROLLARY (MA + ¬CH)

For each free ultrafilter $\mathcal{U}$ on $\mathbb{N}$, $(\ell^\infty/\mathcal{U},<_{\mathcal{U}})$ is not a β_1-set.

Proof

By 6.24, there is an embedding $(\mathbb{R},<) \to (\langle Z\rangle/\mathcal{U},<_{\mathcal{U}})$. The map $(\alpha_n) \mapsto (1/\alpha_n)$ induces an anti-isotonic map from $(\langle Z\rangle/\mathcal{U},<_{\mathcal{U}})$ into $(\ell^\infty/\mathcal{U},<_{\mathcal{U}})$, and so there is an embedding of

$(\mathbb{R},<^*)$, the set $\mathbb{R}$ with the reversed order, into ℓ^∞/U. Since $\mathbb{R}$ is not a β_1-set, $(\mathbb{R},<^*)$ is not a β_1-set, and so ℓ^∞/U is not a β_1-set. ∎

Let U and V be free ultrafilters on $\mathbb{N}$. We stated in Chapter 3 (page 50) that, with CH, the algebras $(\ell^\infty/U)^*$, $(\ell^\infty/V)^*$, c_o/U, and c_o/V are all isomorphic as real algebras. We can now give a justification of this remark. In fact, each of these algebras is a totally ordered, real-closed, η_1-field of cardinality $2^{\aleph_o}$. Now any pair of totally ordered, real-closed, η_1-fields of cardinality $\aleph_1$ are isomorphic as real algebras: that they are isomorphic as fields is proved in [36, 13.13], and the extension to the stated result is due to Johnson [47]. Thus, with CH, the result follows.

In view of the Questions raised in Chapter 3, it is of interest to consider when these algebras are isomorphic if CH is not assumed. In fact, it follows from a result of Dow ([22, Corollary 2.3]) that, if $\aleph_1 < 2^{\aleph_o}$, then there exist free ultrafilters U and V on $\mathbb{N}$ such that $(\ell^\infty/U)^*$ and $(\ell^\infty/V)^*$ are not isomorphic (as rings).

We mention one further open question that interests us. We stated in Theorem 1.10 that, with CH, there is a discontinuous homomorphism from $C(X,\mathbb{C})$ into a Banach algebra for each infinite compact space X, and we explained that we shall eventually prove that NDH is relatively consistent with ZFC. But this leaves open the possibility that there are models of ZFC + ¬CH in which there are discontinuous homomorphisms from $C(X,\mathbb{C})$.

QUESTION. *Is* ¬NDH *relatively consistent with* ZFC + ¬CH?

An approach to this question is given in [17]. Let $(K,<)$ be a real ordered field. Then $(K,<)$ is an <u>α_1-field</u>

if it is an α_1-set, and $(K,<)$ is a <u>β_1-field</u> if there is an ordinal σ such that $K = \bigcup\{K_\nu : \nu < \sigma\}$, where $\{K_\nu : \nu < \sigma\}$ is a chain of real α_1-fields in $(P(K),\subset)$. Certainly each field of cardinality $\aleph_1$ is a β_1-field.

The version of Theorem 1.10 actually proved by Dales and Esterle in [17, Chapter 6] implies the following result. Let K be a real algebra such that $(K,<)$ is a β_1-field. Then the real algebra K^* (the set of infinitesimals in K) is normable. We stress that this is a theorem of ZFC. It implies that, if K is a real field which has a transcendence basis over $\mathbb{R}$ of cardinality $\aleph_1$, then K^* is normable: the only rôle of CH in the proof of Theorem 1.9 is to ensure that $C(X)/P$ satisfies this condition.

It follows from the theorem of Dales and Esterle that to answer the above question positively, it suffices to prove that *it is relatively consistent with* ZFC + $\neg$CH *that there is a free ultrafilter* $\mathcal{U}$ *on* $\mathbb{N}$ *such that* $(\mathbb{R}^{\mathbb{N}}/\mathcal{U},<_{\mathcal{U}})$ *is a* β_1*-field.* By 6.28, to achieve this we must work with models in which MA is false. It is proved in [72] that it is relatively consistent with ZFC + $\neg$CH that there is a free ultrafilter $\mathcal{U}$ on $\mathbb{N}$ such that $(\mathbb{R}^{\mathbb{N}}/\mathcal{U},<_{\mathcal{U}})$ is a β_1-set, and it may be that already this example is a β_1-field - but this we cannot see.

Let X be a compact space, and let P be a non-maximal, prime ideal in $C(X,\mathbb{C})$. Even in the theory ZFC + CH, there is an open question about the normability of $C(X,\mathbb{C})/P$.

QUESTION. *If* $C(X,\mathbb{C})/P$ *is normable, is* $|C(X,\mathbb{C})/P| = 2^{\aleph_0}$?

The question generalizes to ordered fields. Let K be an ordered field over $\mathbb{R}$ such that K^* is normable. Is $|K^*| = 2^{\aleph_0}$? For $a,b \in K^*$, set $a \sim b$ if there exists $n \in \mathbb{N}$ with $|a| \leq n|b|$ and $|b| \leq n|a|$. Then $\sim$ is an equivalence relation on K^*: the equivalence classes are the <u>Archimedean classes</u>. Now suppose that K^* is normable.

Since K^* is a radical algebra, the map $\theta : a \mapsto (\|a^n\|)$, $K^* \to c_o$, has the property that $\theta(a) = \theta(b)$ only if $a \sim b$, and so $|K^*/\sim| \leq |c_o| = 2^{\aleph_o}$. This in turn implies that $|K^*| \leq 2^{\mathfrak{c}}$. Assuming CH, it can be shown that there is an ordered field K over $\mathbb{R}$ with $|K^*/\sim| \leq 2^{\aleph_o}$ and $|K^*| = 2^{\aleph_1}$.

Let P be a non-maximal, prime ideal in $C(X)$. Then $C(X)/P$ is an ordered integral domain with respect to the order induced by the standard order on $C(X)$, and its quotient field K_P is an ordered field over $\mathbb{R}$. The fields K_P have additional structure, and so one might hope to prove that $|K_P{}^*| = 2^{\aleph_o}$ if $|K_P{}^*/\sim| = 2^{\aleph_o}$. However, we can show that, assuming the consistency of certain large cardinal axioms, it is consistent with CH that there exist such X and P with $|K_P{}^*| = 2^{\aleph_1}$ and $|K_P{}^*/\sim| = 2^{\aleph_o}$.

6.30 NOTES

Proposition 6.1 is given in [34,12H] and in [49].

The theory of η_1-sets goes back to Hausdorff in 1908. For further theory of η_1-structures, see [17, §1.6], [36, Chapter 13], [54, Chapter 4], and [56, Chapter 9, §4].

Hausdorff's theorem on the existence of $(\aleph_1, \aleph_1)$-gaps in $(\mathcal{P}(\mathbb{N}), <_F)$ was given in [40]; proofs are also given in [14,14.14] and [34,21L].

The notion of a Hausdorff gap and the proof of 6.9 in ZFC are due to Kunen; see also [72]. Theorems 6.12 and 6.13 are also variants of unpublished results of Kunen, and Theorems 6.15 and 6.16 are given in [72]. Proofs of some of the results on gaps are also given in [45,II.§6].

The definitions of α_1- and β_1-sets are given explicitly in [17,§1.6], but the idea is older; see [36, Chapter 13], for example. The set $\mathbb{Q}$ was introduced by Sierpiński, and $\mathbb{R}$ and $\mathbb{Q}$ are discussed in [36, Chapter 13]. It was shown in 6.28(i) that $\mathbb{Q}$ is a totally ordered β_1-η_1-set: in fact there is really only one such set because each totally ordered β_1-η_1-set is order-isomorphic to $(\mathbb{Q},<)$, ([17,1.6.9(i)]).

7 FORCING

We shall now give our presentation of the technique of forcing.

Let ϕ be a sentence of the language $\mathcal{L}$. We explained in Chapter 4 that the algebraic approach to a proof that ϕ is independent of ZFC involves the construction of models, and that forcing is a technique for building extensions of a given model of ZFC. The main result of the present chapter is Theorem 7.34: if there is a model $\mathfrak{M}$ of ZFC, then there is a model, extending $\mathfrak{M}$, of ZFC + ¬CH. We shall also prove, in Theorem 7.38 that, if there is a model $\mathfrak{M}$ of ZFC, then there is a model, extending $\mathfrak{M}$, of ZFC + CH. In conjunction with Gödel's completeness theorem, Theorem 4.8, these results establish that Con ZFC implies Con(ZFC + CH) and Con(ZFC + ¬CH), and that the Continuum Hypothesis CH is independent of the theory ZFC.

There are a number of different ways of presenting forcing (the methods are surveyed in [49, VII, §9]): our presentation is based on Boolean-valued models. The elementary theory of Boolean algebras that we need was given in Chapter 2. However, there is a point of notation that we must explain. Formally, a Boolean algebra is a sextuple

$$\mathfrak{B} = (B, \wedge, \vee, ', 0, 1)$$

which satisfies certain rules (see Definition 2.6). In these last two chapters we shall distinguish between the Boolean algebra (denoted by a fraktur letter) and its underlying set (denoted by the corresponding roman letter).

7.1 DEFINITION

Let $\mathcal{B} = (B, \wedge, \vee, ', 0, 1)$ be a complete Boolean algebra. A <u>$\mathcal{B}$-valued model</u> is a triple $(M, E_{\mathcal{B}}, \sim_{\mathcal{B}})$, where M is a non-empty set and $E_{\mathcal{B}}: (a,b) \mapsto [\![a \mathrel{\hat{\in}} b]\!]$ and $\sim_{\mathcal{B}} : (a,b) \mapsto [\![a \mathrel{\hat{=}} b]\!]$ are maps from $M \times M$ into B such that, for each $a,b,c \in B$:

(i) $[\![a \mathrel{\hat{=}} a]\!] = 1$;

(ii) $[\![a \mathrel{\hat{=}} b]\!] = [\![b \mathrel{\hat{=}} a]\!]$;

(iii) $[\![a \mathrel{\hat{=}} b]\!] \wedge [\![b \mathrel{\hat{=}} c]\!] \leqslant [\![a \mathrel{\hat{=}} c]\!]$;

(iv) $[\![a \mathrel{\hat{=}} b]\!] \wedge [\![b \mathrel{\hat{\in}} c]\!] \leqslant [\![a \mathrel{\hat{\in}} c]\!]$;

(v) $[\![a \mathrel{\hat{=}} b]\!] \wedge [\![c \mathrel{\hat{\in}} b]\!] \leqslant [\![c \mathrel{\hat{\in}} a]\!]$.

The set M is the <u>underlying set</u> of the model, and the model is <u>countable</u> if M is countable.

We shall denote a $\mathcal{B}$-valued model by $\mathfrak{M}_{\mathcal{B}}$, with the convention that $\mathfrak{M}_{\mathcal{B}} = (M, E_{\mathcal{B}}, \sim_{\mathcal{B}})$.

The language $\mathscr{L}$ has two binary relation-symbols, $\hat{\in}$ and $\hat{=}$. In some ways, it would be more natural to define a model to be a triple $(M, E, \sim)$, where M is a non-empty set and E and $\sim$ are binary relations on M, for this would produce a symmetric treatment of $\hat{\in}$ and $\hat{=}$. Since $\hat{=}$ is to represent "equality", we would require that $\sim$ is an <u>equivalence relation that E respects</u>, in the sense that, for all $a,b,c \in M$:

(i) $a \sim a$;

(ii) if $a \sim b$, then $b \sim a$;

(iii) if $a \sim b$ and $b \sim c$, then $a \sim c$;

(iv) if $a \sim b$ and bEc, then aEc;

(v) if $a \sim b$ and cEb, then cEa.

If $\sim$ is such a relation, the binary relation $E/\sim$ is well defined, and $(M/\sim, E/\sim)$ is a model.

A model $(M, E, \sim)$ in the above sense can clearly be regarded as a $\mathcal{B}$-valued model for each complete Boolean algebra $\mathcal{B}$: set $E_{\mathcal{B}}(a,b) = 1$ if aEb, $E_{\mathcal{B}}(a,b) = 0$ if $a\not{E}b$, $\sim_{\mathcal{B}}(a,b) = 1$ if $a \sim b$, $\sim_{\mathcal{B}}(a,b) = 0$ if $a \not\sim b$. Thus the concept of a $\mathcal{B}$-valued model is a natural generalization of that of a model: in a $\mathcal{B}$-valued model, we can consider sentences whose "truth value" lies strictly between 0 and 1.

Let $\mathscr{L}$ be the language $\mathscr{L}(\hat{\epsilon},\hat{=})$, as in Chapter 4, and let $\phi(x_1,\ldots,x_n)$ be a formula of $\mathscr{L}$. In Chapter 4 (pages 61/62), we defined the truth of ϕ at $(a_1,\ldots,a_n)$ in a model $\mathfrak{M}$. Now let $\mathfrak{M}_{\mathcal{B}}$ be a $\mathcal{B}$-valued model, and let $a_1,\ldots,a_n \in M$. We shall define the <u>truth of ϕ at $(a_1,\ldots,a_n)$ in</u> $\mathfrak{M}_{\mathcal{B}}$ to be a certain element of B, to be denoted by

$$[\![\phi[a_1,\ldots,a_n]]\!]^{\mathfrak{M}_{\mathcal{B}}} \quad \text{or} \quad [\![\phi[a_1,\ldots,a_n]]\!].$$

For example, if ϕ is the atomic formula '$x_1 \hat{\epsilon}\, x_2$', it is natural to define $[\![\phi[a_1,a_2]]\!]$ to be $E_{\mathcal{B}}(a_1,a_2)$.

Formally, we define $[\![\phi[a_1,\ldots,a_n]]\!]$ by rules to be set out below. The rules proceed by induction on the length of formulae, and are essentially the same as those given in Chapter 4.

1 Let x_i and x_j be variables. Then:

$$[\![(x_i \,\hat{=}\, x_j)[a_1,\ldots,a_n]]\!] = \sim_{\mathcal{B}}(a_i,a_j);$$

$$[\![(x_i \,\hat{\epsilon}\, x_j)[a_1,\ldots,a_n]]\!] = E_{\mathcal{B}}(a_i,a_j).$$

2 Let ϕ and ψ be formulae. Then:

$$[\![(\phi \vee \psi)[a_1,\ldots,a_n]]\!] = [\![\phi[a_1,\ldots,a_n]]\!] \vee [\![\psi[a_1,\ldots,a_n]]\!];$$

$$[\![(\neg\phi)[a_1,\ldots,a_n]]\!] = [\![\phi[a_1,\ldots,a_n]]\!]'.$$

3 Let ϕ be a formula, and let x_i be a variable.

Then:

$$[\![(\exists x_i \phi)[a_1,\ldots,a_n]]\!] =$$

$$\bigvee\{ [\![\phi[a_1,\ldots,a_{i-1},b,a_{i+1},\ldots,a_n]]\!] : b \in M\};$$

Note that it is because the Boolean algebra $\mathcal{B}$ is complete that the suprema arising in Rule 3 exist in B.

The above definition of $[\![\phi[a_1,\ldots,a_n]]\!]$ again implicitly assumes that all the variables, bound or free, occurring in ϕ belong to the set $\{x_1,\ldots,x_n\}$. If x_i is a bound variable of ϕ, then $[\![\phi[a_1,\ldots,a_n]]\!]$ does not depend on a_i, and so $[\![\phi[a_1,\ldots,a_n]]\!]$ is well defined whenever the free variables of ϕ belong to $\{x_1,\ldots,x_n\}$. In particular, $[\![\phi]\!]$ is a well-defined element of B for each sentence ϕ.

It follows from the definitions on page 56 that, if ϕ and ψ are formulae, then

$$[\![(\phi \wedge \psi)[a_1,\ldots,a_n]]\!] = [\![\phi[a_1,\ldots,a_n]]\!] \wedge [\![\psi[a_1,\ldots,a_n]]\!],$$

and that, if ϕ is a formula and x_i is a variable, then

$$[\![(\forall x_i \phi)[a_1,\ldots,a_n]]\!] =$$

$$\bigwedge\{ [\![\phi[a_1,\ldots,a_{i-1},b,a_{i+1},\ldots,a_n]]\!] : b \in M\}.$$

Certainly, if $\mathfrak{M}$ is a model, then $\mathfrak{M} \models \phi[a_1,\ldots,a_n]$ if and only if $[\![\phi[a_1,\ldots,a_n]]\!]^{\mathfrak{M}} = 1$ when we regard $\mathfrak{M}$ as a $\mathcal{B}$-valued model.

Let T be a non-empty theory. Then we set

$$[\![T]\!] = \bigwedge\{ [\![\phi]\!] : \phi \in T\}.$$

We shall use ultrafilters to produce models from Boolean-valued models.

Let $\mathfrak{M}_{\mathcal{B}}$ be a $\mathcal{B}$-valued model, and let $\mathcal{U}$ be an ultrafilter in $\mathcal{B}$. For $a,b \in M$, set aEb if $E_{\mathcal{B}}(a,b) \in \mathcal{U}$ and set $a \sim b$ if $\sim_{\mathcal{B}}(a,b) \in \mathcal{U}$. Conditions (i) - (v) in Definition 7.1 show that $\sim$ is an equivalence relation that E respects, and so $(M/\sim, E/\sim)$ is a model : it is denoted by $\mathfrak{M}_{\mathcal{B}}(\mathcal{U})$.

7.2 EXAMPLE

Let M be a non-empty subset of $\mathbb{R}^{\mathbb{N}}$, let $\mathcal{B} = P(\mathbb{N})$, and, for $f,g \in M$, set

$$E_{\mathcal{B}}(f,g) = \{n \in \mathbb{N} : f(n) < g(n)\},$$
$$\sim_{\mathcal{B}}(f,g) = \{n \in \mathbb{N} : f(n) = g(n)\}.$$

Then $\mathfrak{M}_{\mathcal{B}} = (M,E_{\mathcal{B}},\sim_{\mathcal{B}})$ is a $\mathcal{B}$-valued model, and, for each ultrafilter $\mathcal{U}$ on $\mathbb{N}$, $\mathfrak{M}_{\mathcal{B}}(\mathcal{U})$ is simply the ordered set $(M/\mathcal{U},<_{\mathcal{U}})$. In particular, if $M = \mathbb{R}^{\mathbb{N}}$, then $\mathfrak{M}_{\mathcal{B}}(\mathcal{U})$ is the ultraproduct of $\mathbb{R}$ by $\mathcal{U}$ (regarded as an ordered set).

Let $\mathfrak{M}_{\mathcal{B}}$ be a $\mathcal{B}$-valued model, let $\phi(x_1,\ldots,x_n)$ be a formula, and let $\mathcal{U}$ be an ultrafilter in $\mathcal{B}$. Consider the following condition:

$$\left.\begin{array}{l}\text{for each } a_1,\ldots,a_n \in M,\ [\![\phi[a_1,\ldots,a_n]]\!] \in \mathcal{U}\\ \text{if and only if } \mathfrak{M}_{\mathcal{B}}(\mathcal{U}) \models \phi[[a_1]_{\mathcal{U}},\ldots,[a_n]_{\mathcal{U}}].\end{array}\right\}(1)$$

It is immediate from the definitions that (1) holds for each ultrafilter $\mathcal{U}$ whenever ϕ is an atomic formula, and one is tempted to try and prove that (1) holds for each formula ϕ and each ultrafilter $\mathcal{U}$. However the result is not true in this generality: there are problems with quantifiers.

For example, consider the case where $M = c_{oo}$, the space of sequences which are eventually zero, in Example 7.2, and let ϕ be the sentence '$\forall x_1 \forall x_2 (x_1 \hat{=} x_2)$'. Then

$$[\![\phi]\!] = \bigwedge\{\{n : f(n) = g(n)\} : f,g,\in c_{oo}\} = \emptyset,$$

and so, for each ultrafilter $\mathcal{U}$ on $\mathbb{N}$, $[\![\phi]\!] \notin \mathcal{U}$. However, if $\mathcal{U}$ is a free ultrafilter, then $f =_{\mathcal{U}} g$ for each $f, g \in c_{oo}$, and so $\mathfrak{M}_{\mathcal{B}}(\mathcal{U}) \models \phi$.

We obtain the condition (1) for each formula ϕ and each ultrafilter $\mathcal{U}$ by considering only a special type of $\mathcal{B}$-valued model.

7.3 DEFINITION

Let $\mathfrak{M}_{\mathcal{B}}$ be a $\mathcal{B}$-valued model. Then $\mathfrak{M}_{\mathcal{B}}$ is <u>full</u> if, for each formula $\phi(x_1,\ldots,x_n)$ and each $a_1,\ldots,a_n \in M$, there exists $a \in M$ such that

$$[\![(\exists x_i\phi)[a_1,\ldots,a_n]]\!] = [\![\phi[a_1,\ldots,a_{i-1},a,a_{i-1},\ldots,a_n]]\!].$$

7.4 PROPOSITION

Let $\mathfrak{M}_{\mathcal{B}}$ be a full $\mathcal{B}$-valued model, and let $\mathcal{U}$ be an ultrafilter in $\mathcal{B}$. Then condition (1) is satisfied for each formula ϕ. In particular, if ϕ is a sentence, then $\mathfrak{M}_{\mathcal{B}}(\mathcal{U}) \models \phi$ if and only if $[\![\phi]\!] \in \mathcal{U}$.

<u>Proof</u>

Let Φ be the set of formulae for which (1) is satisfied. Then Φ contains all atomic formulae. Suppose that Φ contains ϕ. Since $\mathcal{U}$ is an ultrafilter, $[\![(\neg\psi)[a_1,\ldots,a_n]]\!] \in \mathcal{U}$ if and only if $[\![\psi[a_1,\ldots,a_n]]\!] \notin \mathcal{U}$, and so $\neg\phi$ belongs to Φ. Similarly, Φ contains $\phi \vee \psi$ whenever Φ contains ϕ and ψ. Since $\mathfrak{M}_{\mathcal{B}}$ is full, Φ contains $\exists x_i\phi$ for each variable x_i whenever ϕ belongs to Φ.

It follows that $\Phi = \mathscr{L}$, and so (1) holds for each formula ϕ. ∎

For example, let $\mathfrak{M}_{\mathcal{B}} = (\mathbb{N}^{\mathbb{N}}, E_{\mathcal{B}}, \sim_{\mathcal{B}})$, as in Example 7.2. We *claim* that, for each formula $\phi(x_1,\ldots,x_n)$ and each $f_1,\ldots,f_n \in \mathbb{N}^{\mathbb{N}}$,

$$[\![\phi[f_1,\dots,f_n]]\!] = \{k \in \mathbb{N} : (\mathbb{N},<) \models \phi[f_1(k),\dots,f_n(k)]\}.$$

It follows from this claim that $\mathfrak{M}_{\mathcal{B}}$ is full: to find a suitable element $f \in \mathbb{N}^{\mathbb{N}}$, it suffices to choose a suitable $f(k)$ for each $k \in \mathbb{N}$. The details of this example are left as an (unimportant) exercise.

7.5 PROPOSITION

Let $\mathfrak{M}_{\mathcal{B}}$ be a $\mathcal{B}$-valued model and let T be a theory with $[\![T]\!] = 1$. Suppose that ϕ is a sentence and that $T \vdash \phi$. Then $[\![\phi]\!] = 1$.

<u>Proof</u>

We can suppose that there is a proof of ϕ from T containing only sentences ([23,§2.4]). Let ψ be a sentence. If ψ is a logical axiom, then it is easily verified (given the list of logical axioms) that $[\![\psi]\!] = 1$. If $\psi \in T$ then $[\![\psi]\!] = 1$ by hypothesis. It remains to see that the rule of inference, Modus Ponens, preserves $[\![\dots]\!] = 1$. For this suppose that ψ and χ are sentences with $[\![\psi]\!] = [\![\psi \rightarrow \chi]\!] = 1$. Then $[\![\chi]\!] = [\![\psi]\!]' \vee [\![\chi]\!] = [\![(\neg\psi) \vee \chi]\!] = [\![\psi \rightarrow \chi]\!] = 1$.

More formally, let $(\phi_1,\dots,\phi_n)$ be a proof of ϕ from T containing only sentences. Thus $\phi_n = \phi$ and for each $i < n$, ϕ_i is a logical axiom, an element of T or obtained from the set $\{\phi_1,\dots,\phi_{i-1}\}$ by Modus Ponens. It now follows by induction on i that for each $i \leqslant n$, $[\![\phi_i]\!] = 1$ and so $[\![\phi]\!] = 1$. ∎

We can now give some more explanation of how, assuming that we are given a model of ZFC, we shall build models of ZFC + ¬CH.

Let $\mathfrak{M} = (M,E)$ be a model of ZFC. We shall select a certain complete Boolean algebra $\mathcal{B}$ and construct a full $\mathcal{B}$-valued model $\mathfrak{M}_{\mathcal{B}}$ such that $[\![\phi]\!] = 1$ for each axiom ϕ of ZFC (so that, by 7.5, we have $[\![\text{ZFC}]\!] = 1$), whereas $[\![\text{CH}]\!] = 0$ (and so $[\![\neg\text{CH}]\!] = 1$). By 7.4, $\mathfrak{M}_{\mathbf{B}}(\mathcal{U}) \models \text{ZFC} + \neg\text{CH}$ for each ultrafilter $\mathcal{U}$ in $\mathbf{B}$, and so $\mathfrak{M}_{\mathbf{B}}(\mathcal{U})$ will be the required model. The construction of $\mathfrak{M}_{\mathbf{B}}$ will be accomplished by working "within $\mathfrak{M}$". For this construction, we need the notion of a "Boolean-valued universe $V^{\mathbf{B}}$", and so this is our next topic. The universe $V^{\mathbf{B}}$ should be compared with the cumulative hierarchy of sets, defined in Chapter 5.

Recall that we write dom f and ran f for the domain and range of a function f, respectively.

7.6 DEFINITION

Let $\mathbf{B}$ be a complete Boolean algebra. We define by recursion on Ord a set $V^{\mathbf{B}}_{\alpha}$ for each $\alpha \in$ Ord:

$$V^{\mathbf{B}}_{0} = \emptyset;$$
$$V^{\mathbf{B}}_{\alpha+1} = \{u : \text{dom } u \subset V^{\mathbf{B}}_{\alpha} \text{ and } \text{ran } u \subset B\};$$
$$V^{\mathbf{B}}_{\alpha} = \bigcup\{V^{\mathbf{B}}_{\beta} : \beta < \alpha\} \text{ for a limit ordinal } \alpha.$$

The class obtained,

$$V^{\mathbf{B}} = \bigcup\{V^{\mathbf{B}}_{\alpha} : \alpha \in \text{Ord}\},$$

is the $\mathbf{B}$-valued universe. The members of $V^{\mathbf{B}}$ are terms.

We regard $\emptyset$ as a function with domain $\emptyset$, and so $V^{\mathbf{B}}_{1} = \{\emptyset\} = V_1$. For example,

$$V^{\mathbf{B}}_{2} = \{\emptyset\} \cup \{\{(\emptyset,b)\} : b \in B\}.$$

It is immediate by induction on α that $V^{\mathbf{B}}_{\alpha} \subset V^{\mathbf{B}}_{\alpha+1}$, and $V^{\mathbf{B}}_{\alpha} \neq V^{\mathbf{B}}_{\alpha+1}$ because $|V^{\mathbf{B}}_{\alpha+1}| \geq |P(V^{\mathbf{B}}_{\alpha})| > |V^{\mathbf{B}}_{\alpha}|$.

For each $u \in V^{\mathbf{B}}$, the rank of u, $\rho(u)$, is the minimum ordinal α such that $u \in V^{\mathbf{B}}_{\alpha+1}$. If $v \in \text{dom } u$, then $\rho(v) < \rho(u)$, and $\rho(u) = 0$ if and only if $u = \emptyset$.

Let the trivial Boolean algebra be (2), and define $\pi : V^{(2)} \to V$ inductively by setting $v(\emptyset) = \emptyset$ and $\pi(v) = \{\pi(v(u)) : u \in \text{dom } v \text{ and } v(u) = 1\}$. Then π is a surjection and $\text{rk}(\pi(u)) \leq \rho(u)$ $(u \in V^{(2)})$, where rk denotes the rank of a set in V.

We shall require an ordering on $\underset{\sim}{\text{Ord}}^n$: set $(\alpha_1,\ldots,\alpha_n) < (\beta_1,\ldots,\beta_n)$ if $\max\{\alpha_1,\ldots,\alpha_n\} < \max\{\beta_1,\ldots,\beta_n\}$ or if $\max\{\alpha_1,\ldots,\alpha_n\} = \max\{\beta_1,\ldots,\beta_n\}$ and $(\alpha_1,\ldots,\alpha_n) \prec (\beta_1,\ldots,\beta_n)$ in the lexicographic ordering $\prec$ on $\underset{\sim}{\text{Ord}}^n$. The ordering $<$ is a well-ordering on $\underset{\sim}{\text{Ord}}^n$, called the <u>canonical well-ordering</u>. For example, if $w \in \text{dom } u$, then $(\rho(w),\rho(v)) < (\rho(u),\rho(v))$, $(\rho(v),\rho(w)) < (\rho(v),\rho(u))$, *and* $(\rho(w),\rho(v)) < (\rho(v),\rho(u))$ in $\underset{\sim}{\text{Ord}}^2$. Each initial segment of $(\underset{\sim}{\text{Ord}}^n,<)$ is a set (but this is not true for $(\underset{\sim}{\text{Ord}}^n,\prec)$, which explains why we do not use $\prec$ itself).

Let u and v be terms. We shall simultaneously define $[\![u \mathbin{\hat{\in}} v]\!]$ and $[\![u \mathbin{\hat{=}} v]\!]$ as elements of B. The definition proceeds by recursion on the position of $(\rho(u),\rho(v))$ in $(\underset{\sim}{\text{Ord}}^2,<)$. The intermediate element $[\![u \subset v]\!]$ is at present just a notational convenience.

7.7 DEFINITION

Let $\mathbf{B}$ be a complete Boolean algebra. Then:

(i) $[\![\emptyset \mathbin{\hat{=}} \emptyset]\!] = 1$;

(ii) for $u \in V^{\mathbf{B}}$, $[\![u \mathbin{\hat{\in}} \emptyset]\!] = 0$ and $[\![\emptyset \subset u]\!] = 1$;

(iii) for $u \in V^{\mathbf{B}}\backslash\{\emptyset\}$.

$$[\![u \subset \emptyset]\!] = [\![u \mathbin{\hat{=}} \emptyset]\!] = [\![\emptyset \mathbin{\hat{=}} u]\!]$$
$$= \bigwedge\{u(t)' : t \in \text{dom } u\};$$

(iv) for $u \in V^{\mathbf{B}}$ and $v \in V^{\mathbf{B}}\backslash\{\emptyset\}$,

$$[\![u \mathbin{\hat{\in}} v]\!] = \bigvee\{v(t) \wedge [\![u \mathbin{\hat{=}} t]\!] : t \in \text{dom } v\}; \quad (2)$$

(v) for $u,v \in V^{\mathbf{B}}\backslash\{\emptyset\}$,

$$[\![u \subset v]\!] = \bigwedge\{u(t)' \vee [\![t \mathrel{\hat{\in}} v]\!] : t \in \operatorname{dom} u\}, \tag{3}$$

and

$$[\![u \mathrel{\hat{=}} v]\!] = [\![u \subset v]\!] \wedge [\![v \subset u]\!].$$

The formulae (2) and (3) are of the greatest importance, and they will be used, usually implicitly, many times in the future to calculate Boolean values. The reader will find that there is a considerable net saving of effort if they are committed to memory.

We attempt an explanation of the intuition behind Definition 7.7.

First, consider the special case where $\mathbf{B} = (2)$. The natural definition of $[\![u \mathrel{\hat{\in}} v]\!]$ in this case is

$$[\![u \mathrel{\hat{\in}} v]\!] = \left.\begin{array}{ll} 1 & \text{if } \pi(u) \in \pi(v), \\ 0 & \text{if } \pi(u) \notin \pi(v), \end{array}\right\}$$

where π was defined on page 138. This is exactly what our definition gives.

Now consider the general case. First, we want the analogues of 7.1(i) - (v) to hold; for example, we want to have $[\![u \mathrel{\hat{\in}} v]\!] \geqslant [\![u \mathrel{\hat{=}} w]\!] \wedge [\![w \mathrel{\hat{\in}} v]\!]$. Secondly, we want $[\![u \mathrel{\hat{\in}} v]\!]$ to be related to the value $v(u)$ for $u \in \operatorname{dom} v$. A first attempt at the definition of $[\![u \mathrel{\hat{\in}} v]\!]$ is to set $[\![u \mathrel{\hat{\in}} v]\!] = v(u)$ for $u \in \operatorname{dom} v$ (and $[\![u \mathrel{\hat{\in}} v]\!] = 0$ for $u \notin \operatorname{dom} v$). But this definition fails to satisfy our desiderata. For example, it could happen that $v(u) = 0$, but that $v(w) > 0$ for some $w \in \operatorname{dom} u$ with $[\![u \mathrel{\hat{=}} w]\!] = 1$: in this case, we must have $[\![u \mathrel{\hat{\in}} v]\!] \geqslant v(w)$. We are essentially forced by these considerations to our form of (2); a similar justification applies to (3).

We next show that the analogues of 7.1(i) - (v) do hold for $u,v,w \in V^{\mathbf{B}}$. This appears to prove that $(V^{\mathbf{B}}, E_{\mathbf{B}}, \sim_{\mathbf{B}})$

is a $\mathbf{B}$-valued model, where $E_{\mathbf{B}}(u,v) = [\![u \mathrel{\hat{\in}} v]\!]$ and $\sim_{\mathbf{B}}(u,v) = [\![u \mathrel{\hat{=}} v]\!]$, but this is not the case because $V^{\mathbf{B}}$ is a class and not a set.

7.8 PROPOSITION

Let $u,v \in V^{\mathbf{B}}$. Then:

(i) $[\![u \mathrel{\hat{=}} u]\!] = 1$;

(ii) $[\![u \mathrel{\hat{=}} v]\!] = [\![v \mathrel{\hat{=}} u]\!]$.

Proof

(i) If $\rho(u) = 0$, then $[\![u \mathrel{\hat{=}} u]\!] = 1$ by 7.7(i). Now suppose that $\rho(u) > 0$ and that $[\![v \mathrel{\hat{=}} v]\!] = 1$ for all $v \in V^{\mathbf{B}}$ with $\rho(v) < \rho(u)$. For each $t \in \text{dom } u$,

$$[\![t \mathrel{\hat{\in}} u]\!] = \bigvee\{u(s) \wedge [\![t \mathrel{\hat{=}} s]\!] : s \in \text{dom } u\}$$

$$\geqslant u(t) \wedge [\![t \mathrel{\hat{=}} t]\!] = u(t)$$

because $[\![t \mathrel{\hat{=}} t]\!] = 1$, and so

$$u(t)' \vee [\![t \mathrel{\hat{\in}} u]\!] \geqslant u(t)' \vee u(t) = 1.$$

Thus $[\![u \subset u]\!] = 1$, and so $[\![u \mathrel{\hat{=}} u]\!] = 1$. The induction continues.

(ii) This is immediate from the definition in 7.7(v). ∎

7.9 PROPOSITION

Let $u,v,w \in V^{\mathbf{B}}$. Then:

(i) $[\![u \mathrel{\hat{=}} v]\!] \wedge [\![v \mathrel{\hat{=}} w]\!] \leqslant [\![u \mathrel{\hat{=}} w]\!]$;

(ii) $[\![u \mathrel{\hat{=}} v]\!] \wedge [\![v \mathrel{\hat{\in}} w]\!] \leqslant [\![u \mathrel{\hat{\in}} w]\!]$;

(iii) $[\![u \mathrel{\hat{=}} v]\!] \wedge [\![w \mathrel{\hat{\in}} v]\!] \leqslant [\![w \mathrel{\hat{\in}} u]\!]$.

Proof

The proof is by induction on the position of $(\rho(u),\rho(v),\rho(w))$ in $(\underset{\sim}{\text{Ord}}^3,<)$. However, to make the induction work, we must add to the statements (i) - (iii) the 15 additional statements obtained by permuting $u, v,$ and w.

Suppose then that the 18 statements hold for each $r, s, t \in V^{B}$ with $(\rho(r), \rho(s), \rho(t)) < (\rho(u), \rho(v), \rho(w))$ in $(\mathrm{Ord}^3, <)$. We must prove the 18 statements for u, v, w. We shall prove (i), (ii), and (iii): the remaining 15 statements follow by permuting $u, v,$ and w in the argument.

First, we prove that

$$[\![v \hat{=} w]\!] \wedge [\![u \subset v]\!] \leq [\![u \subset w]\!]. \tag{4}$$

This is immediate if $u = \emptyset$, for then $[\![u \subset w]\!] = 1$. Now suppose that $\rho(u) > 0$, and take $t \in \mathrm{dom}\, u$. Then

$$[\![v \hat{=} w]\!] \wedge (u(t)' \vee [\![t \hat{\in} v]\!])$$
$$\leq u(t)' \vee ([\![v \hat{=} w]\!] \wedge [\![t \hat{\in} v]\!]),$$

and, by the inductive hypothesis and 7.8(ii),

$$[\![v \hat{=} w]\!] \wedge [\![t \hat{\in} v]\!] \leq [\![t \hat{\in} w]\!].$$

Thus

$$[\![v \hat{=} w]\!] \wedge (u(t)' \vee [\![t \hat{\in} v]\!])$$
$$\leq u(t)' \vee [\![t \hat{\in} w]\!],$$

and (4) follows.

By symmetry, (4) also holds for any permutation of u, v, w, and, in particular,

$$[\![v \hat{=} u]\!] \wedge [\![w \subset v]\!] \leq [\![w \subset u]\!]. \tag{5}$$

From (4) and (5),

$$([\![v \hat{=} w]\!] \wedge [\![u \subset v]\!]) \wedge ([\![v \hat{=} u]\!] \wedge [\![w \subset v]\!])$$
$$\leq [\![u \subset w]\!] \wedge [\![w \subset u]\!] = [\![u \hat{=} w]\!].$$

Since $[\![u \subset v]\!] \geqslant [\![v \mathrel{\hat{=}} u]\!]$ and $[\![w \subset v]\!] \geqslant [\![v \mathrel{\hat{=}} w]\!]$, condition (i) follows.

Condition (ii) is immediate if $w = \emptyset$, for then $[\![v \mathrel{\hat{\in}} w]\!] = 0$. Now suppose that $\rho(w) > 0$, and take $t \in \operatorname{dom} w$. By the inductive hypothesis,

$$[\![u \mathrel{\hat{=}} v]\!] \wedge [\![v \mathrel{\hat{=}} t]\!] \leqslant [\![u \mathrel{\hat{=}} t]\!],$$

and so

$$[\![u \mathrel{\hat{=}} v]\!] \wedge (w(t) \wedge [\![v \mathrel{\hat{=}} t]\!]) \leqslant w(t) \wedge [\![u \mathrel{\hat{=}} t]\!],$$

an inequality from which condition (ii) follows.

We prove a stronger version of (iii), namely that

$$[\![v \subset u]\!] \wedge [\![w \mathrel{\hat{\in}} v]\!] \leqslant [\![w \mathrel{\hat{\in}} u]\!]. \tag{6}$$

This is immediate if $v = \emptyset$, for then $[\![w \mathrel{\hat{\in}} v]\!] = 0$. Now suppose that $\rho(v) > 0$, and take $t \in \operatorname{dom} v$. Then

$$[\![v \subset u]\!] \leqslant v(t)' \vee [\![t \mathrel{\hat{\in}} u]\!],$$

and so

$$\begin{aligned}[\![v \subset u]\!] &\wedge (v(t) \wedge [\![w \mathrel{\hat{=}} t]\!]) \\ &\leqslant [\![w \mathrel{\hat{=}} t]\!] \wedge [\![t \mathrel{\hat{\in}} u]\!] \leqslant [\![w \mathrel{\hat{\in}} u]\!]\end{aligned}$$

by the inductive hypothesis. This is an inequality from which (6) follows. ∎

There is a further preliminary result that we shall require later.

7.10 LEMMA

For $u \in V^{\mathbf{B}}$ and $a \in B$, define u_a by

$$\operatorname{dom} u_a = \operatorname{dom} u, \quad u_a(t) = u(t) \wedge a \quad (t \in \operatorname{dom} u).$$

Then $u \in V^{\mathcal{B}}$, and $[\![u \mathrel{\hat{=}} u_a]\!] \geq a$.

Proof

Clearly, for $t \in \operatorname{dom} u$, $[\![t \mathrel{\hat{\in}} u]\!] \geq u(t)$ and $[\![t \mathrel{\hat{\in}} u_a]\!] \geq u_a(t)$, and so

$$u(t)' \vee [\![t \mathrel{\hat{\in}} u_a]\!] \geq u(t)' \vee (u(t) \wedge a) \geq a,$$

$$u_a(t)' \vee [\![t \mathrel{\hat{\in}} u]\!] \geq (u(t) \wedge a)' \vee u(t) = 1.$$

Thus $[\![u \subset u_a]\!] \geq a$, $[\![u_a \subset u]\!] = 1$, and $[\![u \mathrel{\hat{=}} u_a]\!] \geq a$. ∎

For $u,v \in V^{\mathcal{B}}$, set $u \sim v$ if $[\![u \mathrel{\hat{=}} v]\!] = 1$. Then $\sim$ is an equivalence relation on $V^{\mathcal{B}}$: u and v are equivalent if $u \sim v$. We write

$$\operatorname{supp} u = \{t \in \operatorname{dom} u : u(t) \neq 0\}.$$

Set $u \approx v$ if $\operatorname{supp} u = \operatorname{supp} v$ and $u(t) = v(t)$ $(t \in \operatorname{supp} u)$. Then $\approx$ is an equivalence relation on $V^{\mathcal{B}}$, and $u \sim v$ if $u \approx v$.

Recall from 5.15(i) that a subset A of a Boolean algebra $\mathcal{B}$ is an antichain if $A \subset B\backslash\{0\}$ and if $a \wedge b = 0$ whenever $a,b \in A$ with $a \neq b$. The following result is sometimes called the mixing lemma.

7.11 PROPOSITION

Let A be an antichain in the complete Boolean algebra $\mathcal{B}$, and let $f : A \to V^{\mathcal{B}}$ be a function. Then there exists $u \in V^{\mathcal{B}}$ with $[\![u \mathrel{\hat{=}} f(a)]\!] \geq a$ $(a \in A)$.

Proof

Set $\operatorname{dom} u = \bigcup\{\operatorname{dom} f(a) : a \in A\}$, and, for $t \in \operatorname{dom} u$, set

$$u(t) = \bigvee\{f(a)(t) \wedge a : a \in A \text{ with } t \in \operatorname{dom} f(a)\}.$$

Take $a \in A$. Since A is an antichain, we have $u(t) \wedge a = f(a)(t) \wedge a$ if $t \in \operatorname{dom} f(a)$, and $u(t) \wedge a = 0$ if $t \notin \operatorname{dom} f(a)$. Now $u_a \approx f(a)_a$, and so $[\![u_a \hat{=} f(a)_a]\!] = 1$. By 7.10, $[\![u \hat{=} u_a]\!] \geq a$ and $[\![f(a) \hat{=} f(a)_a]\!] \geq a$, and so, by 7.9(i), $[\![u \hat{=} f(a)]\!] \geq a$. ∎

Let $\phi(x_1,\ldots,x_n)$ be a formula of $\mathscr{L}$, and let $u_1,\ldots,u_n \in V^{\mathbf{B}}$. We define the <u>truth of</u> ϕ <u>at</u> $(u_1,\ldots,u_n)$ <u>in</u> $V^{\mathbf{B}}$ to be an element of B, to be again denoted by

$$[\![\phi[u_1,\ldots,u_n]]\!],$$

as if $(V^{\mathbf{B}},E_{\mathbf{B}},\sim_{\mathbf{B}})$ were a $\mathbf{B}$-valued model. Formally, the rules are exactly those set out on pages 132/133. Some care is required with Rule 3, however. The problem is that we must be sure that each supremum of the form $\bigvee\{\ldots : u \in V^{\mathbf{B}}\}$ exists in B: we are apparently taking a supremum over a proper class. To show the suprema do exist, it is sufficient to show that

$$\{a \in B : a = [\![\phi[u_1,\ldots,u_{i-1},u,u_{i+1},\ldots,u_n]]\!] \text{ for some } u \in V^{\mathbf{B}}\},$$

is actually a set. We shall note on page 156 that, for each formula $\psi(x_1,\ldots,x_m)$, there is a formula $\psi^*(x_1,\ldots,x_{m+2})$ such that, for $v_1,\ldots,v_m \in V^{\mathbf{B}}$, $[\![\psi[v_1,\ldots,v_m]]\!] = a$ in B if and only if $\psi^*[v_1,\ldots,v_m,a,B]$ is true in V, and so the result follows from the Axiom of Comprehension because B is a set. (The formula ψ^* does not assert that $[\![\psi[v_1,\ldots,v_m]]\!]$ exists, but it specifies the conditions on $a \in B$ that ensure that $a = [\![\psi[v_1,\ldots,v_m]]\!]$. The construction of ψ^* depends on both ψ and m, and involves working through the finitely many subformulae of ψ.)

Since this is a book about forcing, we should at least mention the standard notation used in the subject. Let

$\mathcal{B}$ be a complete Boolean algebra, and let $b \in B$. Then one writes

$$b \Vdash \text{'}\phi[u_1,\ldots,u_n]\text{'} \quad \text{for} \quad b \leqslant [\![\phi[u_1,\ldots,u_n]]\!] \quad \text{in} \quad V^{\mathcal{B}}.$$

However, we shall not use this notation.

On page 65, we introduced the abbreviation $x \subset y$ for $\forall t(t \mathrel{\hat{\in}} x \rightarrow t \mathrel{\hat{\in}} y)$. We now see that this is consistent with the notation in Definition 7.7. We must show that, for each $u, v \in V^{\mathcal{B}}$,

$$[\![u \subset v]\!] = \bigwedge\{[\![w \mathrel{\hat{\in}} u]\!]' \vee [\![w \mathrel{\hat{\in}} v]\!] : w \in V^{\mathcal{B}}\}. \quad (7)$$

This is immediate if $u = \emptyset$. Now suppose that $\rho(u) > 0$. By (6), $[\![u \subset v]\!] \wedge [\![w \mathrel{\hat{\in}} u]\!] \leqslant [\![w \mathrel{\hat{\in}} v]\!]$, and so $[\![u \subset v]\!] \leqslant [\![w \mathrel{\hat{\in}} u]\!]' \vee [\![w \mathrel{\hat{\in}} v]\!]$ for $w \in V^{\mathcal{B}}$. Also, for $t \in \text{dom } u$, $u(t) \leqslant [\![t \mathrel{\hat{\in}} u]\!]$, and so

$$[\![u \subset v]\!] \geqslant \bigwedge\{[\![t \mathrel{\hat{\in}} u]\!]' \vee [\![t \mathrel{\hat{\in}} v]\!] : t \in \text{dom } u\}.$$

Thus (7) follows.

Observe that $\emptyset \in V^{\mathcal{B}}$, and so $[\![u \mathrel{\hat{=}} \emptyset]\!]$ has two possible interpretations: the first involves the equality of u and $\emptyset$ as terms, the second is $[\![\phi[u]]\!]$ where $\phi(x) = \forall z(z \mathrel{\hat{\notin}} x)$. Fortunately, both interpretations give the same value.

Part (i) of the following result for a formula $\phi(x)$ is

$$[\![(u \mathrel{\hat{=}} v \vee \phi(u)) \rightarrow \phi(v)]\!] = 1.$$

Part (ii) is the <u>fullness lemma</u> for $V^{\mathcal{B}}$: its proof appeals to the Axiom of Choice, and indeed the result can be shown to be equivalent to the Axiom of Choice.

Let $\phi(x_1,\ldots,x_n)$ be a fixed formula of $\mathcal{L}$.

7.12 THEOREM

Let $\mathbf{B}$ be a complete Boolean algebra, and let $u_1,\ldots,u_n \in V^{\mathbf{B}}$.

(i) For each $u \in V^{\mathbf{B}}$,

$$[\![u \mathrel{\hat{=}} u_i]\!] \wedge [\![\phi[u_1,\ldots,u_n]]\!] \leqslant [\![\phi[u_1,\ldots,u_{i-1},u,u_{i+1},\ldots,u_n]]\!].$$

(ii) There exists $u \in V^{\mathbf{B}}$ such that

$$[\![(\exists x_i \phi)[u_1,\ldots,u_n]]\!] = [\![\phi[u_1,\ldots,u_{i-1},u,u_{i+1},\ldots,u_n]]\!].$$

Proof

(i) Let Φ be the set of subformulae ψ of ϕ such that the result holds for ψ.

By 7.9, Φ contains all the atomic formulae. It is an immediate consequence of the definition of truth in $V^{\mathbf{B}}$ that Φ contains $\psi \vee \chi$ whenever Φ contains ψ and χ and $\psi \vee \chi$ is a subformula of ϕ, and that Φ contains $\exists x_j \psi$ for each variable x_j whenever ϕ contains ψ and $\exists x_j \psi$ is a subformula of ϕ.

Suppose that $\psi \in \Phi$ and that $\neg\psi$ is a subformula of ϕ, say $\psi = \psi(x_1)$ for notational convenience. Set $a = [\![u \mathrel{\hat{=}} u_1]\!]$, $b = [\![\psi[u_1]]\!]$, and $c = [\![\psi[u]]\!]$. Then $a \wedge b \leqslant a \wedge c$. Also, $a = [\![u_1 \mathrel{\hat{=}} u]\!]$, and so $a \wedge c \leqslant a \wedge b$. Thus $a \wedge b = a \wedge c$, and so $a \wedge b' = a \wedge c'$. This implies that $\neg\psi \in \Phi$.

It follows that all subformulae ψ of ϕ belong to Φ, and in particular the result holds for ϕ itself.

(ii) Again for the sake of notational sanity, we suppose that $\phi = \phi(x_1)$.

Let $S = \{[\![\phi[u]]\!] : u \in V^{\mathbf{B}}\}$, so that, as before, S is a subset of B. For each $b \in S$, let $f(b)$ be the minimum ordinal σ with $b = [\![\phi[u]]\!]$ for some $u \in V_\sigma^{\mathbf{B}}$. Then $f : S \to \underset{\sim}{\mathrm{Ord}}$ is a function defined by a property, and so, by the Axiom of Replacement, the image of S is a set. So there is an ordinal α with $f(b) \leqslant \alpha \quad (b \in S)$. We have

$$[\![\exists x_1 \phi]\!] = \bigvee\{[\![\phi[u]]\!] : u \in V_\alpha^{\mathbf{B}}\}.$$

Let

$$I = \{b \in B : b \leqslant [\![\phi[u]]\!] \text{ for some } u \in V_\alpha^{\mathbf{B}}\}.$$

By Zorn's Lemma, there is a maximal antichain A in $(I\backslash\{0\},\subset)$, and clearly $[\![\exists x_1 \phi]\!] = \bigvee A$. By the Axiom of Choice again, there is a function $g : A \to V_\alpha^{\mathbf{B}}$ such that $a \leqslant [\![\phi[g(a)]]\!] \quad (a \in A)$, and, by the mixing lemma, there exists $u \in V^{\mathbf{B}}$ with $[\![u \mathrel{\hat{=}} g(a)]\!] \geqslant a \quad (a \in A)$. By (i),

$$[\![u \mathrel{\hat{=}} g(a)]\!] \wedge [\![\phi[g(a)]]\!] \leqslant [\![\phi[u]]\!],$$

and so $[\![\phi[u]]\!] \geqslant a \quad (a \in A)$. Thus $[\![\phi[u]]\!] = [\![\exists x_1 \phi]\!]$. ∎

We shall now show that $[\![\phi]\!] = 1$ for each axiom ϕ of ZFC.

Let us stress that the question whether or not $[\![\phi]\!] = 1$ in $V^{\mathbf{B}}$ for a given sentence ϕ of $\mathscr{L}$ and a complete Boolean algebra $\mathbf{B}$ is a mathematical question, completely analogous to that of determining whether or not ϕ is true in V, and in deciding the question we can argue in ZFC in the normal way.

First we consider the axioms of ZF; the more difficult Axiom of Choice will be dealt with later. It will be seen that to prove that $[\![\phi]\!] = 1$ for an axiom ϕ of ZF we can argue in ZF, but that, naturally enough, the proof that $[\![AC]\!] = 1$ requires AC.

Throughout, we shall use the following facts. Let ψ and χ be statements. Since $\psi \Rightarrow \chi$ is $(\neg\psi) \vee \chi$, we have

$$[\![\psi \Rightarrow \chi]\!] = 1 \text{ if and only if } [\![\psi]\!] \leqslant [\![\chi]\!],$$

$$[\![\psi \Leftrightarrow \chi]\!] = 1 \text{ if and only if } [\![\psi]\!] = [\![\chi]\!].$$

The next result is really an infinite list of theorems, one for each axiom of ZF: it states that certain theorems can be proved, and so it is a metatheorem.

In some cases, we examine an axiom which is logically equivalent to the axiom stated in Chapter 4. That this is sufficient will be proved in 7.16(i) (a result whose proof does not require 7.13).

7.13 METATHEOREM

Let $\mathbb{B}$ be a complete Boolean algebra. Then $[\![\psi]\!] = 1$ in $V^{\mathbb{B}}$ for each axiom ψ of ZF as specified in Chapter 4.

Proof

We shall examine each of the axioms in turn.

1 Extensionality

$$\forall x \forall y((x \hat{=} y) \leftrightarrow \forall z((z \hat{\in} x) \leftrightarrow (z \hat{\in} y)))$$

It is to be proved that, given $u, v \in V^{\mathbb{B}}$,

$$[\![u \hat{=} v]\!] = (\bigwedge\{[\![w \hat{\in} u]\!]' \vee [\![w \hat{\in} v]\!] : w \in V^{\mathbb{B}}\}) \wedge (\bigwedge\{[\![w \hat{\in} v]\!]' \vee [\![w \hat{\in} u]\!] : w \in V^{\mathbb{B}}\}).$$

But this is immediate from (7).

2 Comprehension

Let $\phi(x_1, \ldots, x_n)$ be a formula in $\mathscr{L}$.

$$\forall x_2 \ldots \forall x_n \forall x_{n+1} \exists x_{n+2} \forall x_1((x_1 \hat{\in} x_{n+2}) \leftrightarrow ((x_1 \hat{\in} x_{n+1}) \wedge \phi)$$

It is to be proved that, given $u_2, \ldots, u_n, u_{n+1} \in V^{\mathbb{B}}$, there exists $u_{n+2} \in V^{\mathbb{B}}$ such that, for each $u_1 \in V^{\mathbb{B}}$, $a = b$, where $a = [\![u_1 \hat{\in} u_{n+2}]\!]$ and

$$b = [\![u_1 \mathrel{\hat{\in}} u_{n+1}]\!] \wedge [\![\phi[u_1,\ldots,u_n]]\!].$$

We can suppose that $\rho(u_{n+1}) > 0$. Set $\mathrm{dom}\, u_{n+2} = \mathrm{dom}\, u_{n+1}$, and, for $t \in \mathrm{dom}\, u_{n+2}$, set

$$u_{n+2}(t) = [\![t \mathrel{\hat{\in}} u_{n+1}]\!] \wedge [\![\phi[u_1,\ldots,u_n]]\!].$$

Take $u_1 \in V^{\mathbf{B}}$. Then, for $t \in \mathrm{dom}\, u_{n+2}$,

$$u_{n+2}(t) \wedge [\![u_1 \mathrel{\hat{=}} t]\!] \leqslant [\![u_1 \mathrel{\hat{\in}} u_{n+1}]\!] \wedge [\![\phi[u_1,\ldots,u_n]]\!]$$

by 7.9(ii) and 7.12(i), and so $a \leqslant b$. Also,

$$\begin{aligned} u_{n+1}(t) \wedge [\![u_1 \mathrel{\hat{=}} t]\!] \wedge [\![\phi[u_1,\ldots,u_n]]\!] & \\ \leqslant [\![t \mathrel{\hat{\in}} u_{n+1}]\!] \wedge [\![\phi[u_1,\ldots,u_n]]\!] \wedge [\![u_1 \mathrel{\hat{=}} t]\!] & \\ = u_{n+2}(t) \wedge [\![u_1 \mathrel{\hat{=}} t]\!], & \end{aligned}$$

and so $b \leqslant a$. Thus $a = b$, as required.

3 <u>Pairing</u>

$$\forall x \forall y \exists z ((x \mathrel{\hat{\in}} z) \wedge (y \mathrel{\hat{\in}} z))$$

It is to be proved that, given $u, v \in V^{\mathbf{B}}$, there exists $w \in V^{\mathbf{B}}$ such that $[\![u \mathrel{\hat{\in}} w]\!] \wedge [\![v \mathrel{\hat{\in}} w]\!] = 1$. Set $\mathrm{dom}\, w = \{u, v\}$, $w(u) = w(v) = 1$.

4 <u>Union</u>

$$\forall x \exists y \forall z (z \mathrel{\hat{\in}} x \Rightarrow z \subset y)$$

Given $u \in V^{\mathbf{B}}$, we require $v \in V^{\mathbf{B}}$ such that, for each $w \in V^{\mathbf{B}}$, $[\![w \mathrel{\hat{\in}} u]\!] \leqslant [\![w \subset v]\!]$.

We can suppose that $\rho(u) > 0$. Set

$$\text{dom } v = \bigcup\{\text{dom } t : t \in \text{dom } u\},$$

and, for $s \in \text{dom } v$, set $v(s) = 1$.

For $t \in \text{dom } u$, $[\![t \subset v]\!] = 1$ because $[\![s \hat{\in} v]\!] \geq v(s) \wedge [\![s \hat{=} s]\!] = 1$ for $s \in \text{dom } t$. Take $w \in V^{\mathbf{B}}$. Then

$$u(t) \wedge [\![w \hat{=} t]\!] \leq [\![w \hat{=} t]\!] \wedge [\![t \subset v]\!]$$

$$\leq [\![w \subset v]\!],$$

and so $[\![w \hat{\in} u]\!] \leq [\![w \subset v]\!]$, as required.

5 Power set

$$\forall x \exists y \forall z (z \subset x \Rightarrow z \hat{\in} y)$$

Given $u \in V^{\mathbf{B}}$, we require $v \in V^{\mathbf{B}}$ such that, for each $w \in V^{\mathbf{B}}$, $[\![w \subset u]\!] \leq [\![w \hat{\in} v]\!]$.

We can suppose that $\rho(u) > 0$. Set $\text{dom } v = \mathbf{B}^{\text{dom } u}$, and, for $t \in \text{dom } v$, set $v(t) = 1$. Take $w \in V^{\mathbf{B}}$, and define $f : \text{dom } u \to \mathbf{B}$ by

$$f(t) = [\![t \hat{\in} w]\!] \quad (t \in \text{dom } u).$$

For $t \in \text{dom } f$, $f(t)' \vee [\![t \hat{\in} w]\!] = 1$, and so $[\![f \subset w]\!] = 1$. Also, $f \in \text{dom } v$, and so $[\![f \hat{\in} v]\!] \geq v(f) \wedge [\![f \hat{=} f]\!] = 1$.

For $t \in V^{\mathbf{B}}$,

$$[\![t \hat{\in} w]\!] \wedge [\![t \hat{\in} u]\!]$$

$$\leq \bigvee\{[\![t \hat{\in} w]\!] \wedge [\![s \hat{\in} u]\!] \wedge [\![t \hat{=} s]\!] : s \in \text{dom } u\}$$

$$\leq \bigvee\{[\![s \hat{\in} w]\!] \wedge [\![t \hat{=} s]\!] : s \in \text{dom } u\},$$

and so, for $t \in \text{dom } w$,

$$[\![w \subset u]\!] \leq w(t)' \vee [\![t \mathbin{\hat{\in}} u]\!]$$

$$= w(t)' \vee ([\![t \mathbin{\hat{\in}} w]\!] \wedge [\![t \mathbin{\hat{=}} u]\!])$$

$$\leq w(t)' \vee (\bigvee\{[\![s \mathbin{\hat{\in}} w]\!] \wedge [\![t \mathbin{\hat{=}} s]\!] : s \in \operatorname{dom} u\})$$

$$= w(t)' \vee [\![t \mathbin{\hat{=}} f]\!].$$

Thus $[\![w \subset u]\!] \leq [\![w \subset f]\!]$, and hence

$$[\![w \subset u]\!] \leq [\![w \subset f]\!] \wedge [\![f \subset w]\!] \wedge [\![f \mathbin{\hat{\in}} v]\!]$$

$$\leq [\![w \mathbin{\hat{\in}} v]\!],$$

as required.

6 Replacement

Let $\phi(x_1, x_2)$ be a formula in $\mathscr{L}$.

$$\forall x_3 \exists x_4 \forall x_1 ((x_1 \mathbin{\hat{\in}} x_3 \wedge \exists x_2(\phi)) \Rightarrow \exists x_2 (x_2 \mathbin{\hat{\in}} x_4 \wedge \phi))$$

Given $u_3 \in V^{\mathbf{B}}$, we require $u_4 \in V^{\mathbf{B}}$ such that, for each $u_1 \in V^{\mathbf{B}}$, $[\![u_1 \mathbin{\hat{\in}} u_3]\!] \wedge a \leq b$, where

$$a = [\![(\exists x_2 \phi)[u_1]]\!] = \bigvee\{[\![\phi[u_1, u_2]]\!] : u_2 \in V^{\mathbf{B}}\},$$

$$b = \bigvee\{[\![\phi[u_1, u_2]]\!] \wedge [\![u_2 \mathbin{\hat{\in}} u_4]\!] : u_2 \in V^{\mathbf{B}}\}.$$

By two applications of the Axiom of Replacement, there is an ordinal α such that, for each $v_1 \in \operatorname{dom} u_3$,

$$\bigvee\{[\![\phi[v_1, v_2]]\!] : v_2 \in V^{\mathbf{B}}\} = \bigvee\{[\![\phi[v_1, v_2] : v_2 \in V^{\mathbf{B}}_\alpha\}.$$

Set $\operatorname{dom} u_4 = V^{\mathbf{B}}_\alpha$, and, for $v_2 \in \operatorname{dom} u_4$, set $u_4(v_2) = 1$.

We can suppose that $\rho(u_3) > 0$. For each $v_1 \in \operatorname{dom} u_3$,

$$\llbracket v_1 \hat{=} u_1 \rrbracket \wedge a \leqslant \llbracket v_1 \hat{=} u_1 \rrbracket \wedge \llbracket (\exists x_2 \phi)[v_1] \rrbracket$$

$$= \llbracket v_1 \hat{=} u_1 \rrbracket \wedge \bigvee\{\llbracket \phi[v_1, v_2] \rrbracket : v_2 \in V^{\mathbf{B}}_\alpha\}$$

$$= \bigvee\{\llbracket v_1 \hat{=} u_2 \rrbracket \wedge \llbracket \phi[v_1, v_2] \rrbracket : v_2 \in V^{\mathbf{B}}_\alpha\}$$

$$\leqslant \bigvee\{\llbracket \phi[u_1, u_2] \rrbracket : u_2 \in V^{\mathbf{B}}_\alpha\}.$$

Since $\llbracket u_2 \hat{\in} u_4 \rrbracket = 1$ for $u_2 \in V^{\mathbf{B}}_\alpha$, we have

$$u_3(v_1) \wedge \llbracket v_1 \hat{=} u_1 \rrbracket \wedge a$$

$$\leqslant \bigvee\{\llbracket \phi[u_1, u_2] \rrbracket \wedge \llbracket u_2 \hat{\in} u_4 \rrbracket : u_2 \in V^{\mathbf{B}}_\alpha\},$$

and so $\llbracket u_1 \hat{\in} u_3 \rrbracket \wedge a \leqslant b$, as required.

7 Regularity

This axiom is equivalent to

$$\forall x \exists y (x \hat{=} \emptyset \vee (y \hat{\in} x \wedge (y \cap x \hat{=} \emptyset)))$$

Given $u \in V^{\mathbf{B}}$, we must show that

$$\llbracket u \hat{\neq} \emptyset \rrbracket \leqslant \bigvee\{\llbracket v \hat{\in} u \rrbracket \wedge \llbracket 'u \cap v \hat{=} \emptyset' \rrbracket : v \in V^{\mathbf{B}}\}.$$

Suppose that this is false. Then there exists $b \in B\backslash\{0\}$ such that, for each $v \in V^{\mathbf{B}}$,

$$b \leqslant \llbracket u \hat{\neq} \emptyset \rrbracket, \quad b \wedge \llbracket v \hat{\in} u \rrbracket \wedge \llbracket 'u \cap v \hat{=} \emptyset' \rrbracket = 0.$$

Clearly $u \neq \emptyset$. If $b \wedge \llbracket t \hat{\in} u \rrbracket = 0$ for each $t \in \operatorname{dom} u$, then $b \leqslant u(t)'$ $(t \in \operatorname{dom} u)$, and so $b \leqslant \llbracket u \hat{=} \emptyset \rrbracket$, whence $b = 0$, a contradiction. So we may suppose that $b \wedge \llbracket v \hat{\in} u \rrbracket \neq 0$ for some $v \in V^{\mathbf{B}}$: let α be the minimum ordinal such that $b \wedge \llbracket v \hat{\in} u \rrbracket \neq 0$ for some $v \in V^{\mathbf{B}}_\alpha$.

Let $t \in \operatorname{dom} v$. Then $t \in V^{\mathbf{B}}_\beta$ for some $\beta < \alpha$, and so $b \wedge [\![t \mathrel{\hat{\in}} u]\!] = 0$. But

$$[\![\text{'}u \cap v \mathrel{\hat{\neq}} \emptyset\text{'}]\!] = \bigvee\{ [\![w \mathrel{\hat{\in}} u]\!] \wedge [\![w \mathrel{\hat{\in}} v]\!] : w \in V^{B}\},$$

and

$$[\![w \mathrel{\hat{\in}} u]\!] \wedge [\![w \mathrel{\hat{\in}} v]\!]$$

$$= \bigvee\{ [\![w \mathrel{\hat{\in}} u]\!] \wedge v(t) \wedge [\![w \mathrel{\hat{=}} t]\!] : t \in \mathrm{dom}\ v\}$$

$$\leq \bigvee\{ v(t) \wedge [\![t \mathrel{\hat{\in}} u]\!] : t \in \mathrm{dom}\ v\},$$

so that $b \wedge [\![\text{'}u \cap v \mathrel{\hat{\neq}} \emptyset\text{'}]\!] = 0$. Thus $b \wedge [\![v \mathrel{\hat{\in}} u]\!] =$ $b \wedge [\![v \mathrel{\hat{\in}} u]\!] \wedge [\![\text{'}u \cap v \mathrel{\hat{=}} \emptyset\text{'}]\!] = 0$, a contradiction.

8 Infinity

This axiom is equivalent to

$$\exists x \forall y \exists z (x \mathrel{\hat{\neq}} \emptyset \wedge (y \mathrel{\hat{\in}} x \rightarrow (y \mathrel{\hat{\in}} z \wedge z \mathrel{\hat{\in}} x)))$$

For each $u \in V^{B}$, define $u^* : \{u\} \rightarrow B$ by $u^*(u) = 1$. If $u \in V^{B}_{\alpha}$, then $u^* \in V^{B}_{\alpha+1}$, and, for $u,v \in V^{B}$,

$$[\![u \mathrel{\hat{=}} v]\!] = [\![u^* \mathrel{\hat{=}} v^*]\!] \quad \text{and} \quad [\![u \mathrel{\hat{\in}} u^*]\!] = 1.$$

We require $u \in V^{B}$ such that $[\![u \mathrel{\hat{=}} \emptyset]\!] = 0$ and such that, for each $v \in V^{B}$, there exists $w \in V^{B}$ with $[\![v \mathrel{\hat{\in}} u]\!] \leq [\![v \mathrel{\hat{\in}} w]\!] \wedge [\![w \mathrel{\hat{\in}} u]\!]$.

Set $\mathrm{dom}\ u = V^{B}_{\omega}$, where ω is the first infinite ordinal, and, for $t \in \mathrm{dom}\ u$, set $u(t) = 1$. Then $[\![u \mathrel{\hat{=}} \emptyset]\!] = 0$. Also, for each $v \in V^{B}$,

$$[\![v \mathrel{\hat{\in}} u]\!] = \bigvee\{u(t) \wedge [\![v \mathrel{\hat{=}} t]\!] : t \in \mathrm{dom}\ u\}$$

$$= \bigvee\{ [\![v^* \mathrel{\hat{=}} t^*]\!] : t \in \mathrm{dom}\ u\}$$

$$\leq \bigvee\{ [\![v^* \mathrel{\hat{=}} s]\!] : s \in \mathrm{dom}\ u\}$$

because $t^* \in \operatorname{dom} u$ whenever $t \in \operatorname{dom} u$. Thus

$$[\![v \mathbin{\hat{\in}} u]\!] \leq [\![v^* \mathbin{\hat{\in}} u]\!] = [\![v \mathbin{\hat{\in}} v^*]\!] \wedge [\![v^* \mathbin{\hat{\in}} u]\!]$$

and so u is as required (with $w = v^*$).

This completes the examination of the axioms and the proof of the metatheorem. ∎

To complete the proof that $[\![\mathrm{ZFC}]\!] = 1$ it remains to show that $[\![\mathrm{AC}]\!] = 1$. We postpone this to the end of the chapter, for by then we shall have some more notations and techniques at our disposal. These preliminary results will also be needed in Chapter 8. In fact, when it is slotted in as Theorem 7.41, the proof that $[\![\mathrm{AC}]\!] = 1$ will be gratifyingly brief.

We shall shortly need to consider statements, such as 'x is a choice function for y' or 'x is an ordinal', which are expressed in ordinary mathematical language. For this purpose, we must formalize the statements in the language $\mathscr{L}$.

For example, an ordinal is 'a transitive set which is totally ordered by the relation of membership', and so the statement 'x is an ordinal' can be expressed in $\mathscr{L}$ by

$$'\forall y (y \mathbin{\hat{\in}} x \Rightarrow y \subset x) \wedge$$

$$\wedge \forall z \forall t ((z \mathbin{\hat{\in}} x \wedge t \mathbin{\hat{\in}} x) \Rightarrow (z \mathbin{\hat{=}} t \vee z \mathbin{\hat{\in}} t \vee t \mathbin{\hat{\in}} z))';$$

we regard 'x is an ordinal' as an abbreviation of this formula.

Similarly, we could, under duress, write down the formulae which express 'x is a function from y to z', 'x is a complete Boolean algebra and y is the Stone space of x', 'every homomorphism from ℓ^∞ into a Banach algebra is continuous', etc. The formula expressing this last statement is, of course, a sentence. However, a little experimentation will show that a formula corresponding to even such an innocuous statement as 'x is a function from y to z' is very tedious to write down, and we shall not do so.

We assume that each statement in naïve set theory that we consider can be expressed in the formal language $\mathscr{L}$. Similarly, we assume that, if $\mathfrak{M}$ is a model of ZFC, then, for example,

$$\mathfrak{M} \models \text{'}a \text{ is a function from } b \text{ to } c\text{'}$$

is well defined, where $a,b,c \in M$.

(Some expressions which can be written down cannot be formalized. For example, 'x is a sentence of $\mathscr{L}$ and $[\![x]\!] = 1$ in V^Y' cannot be formalized, even though, for each fixed sentence ϕ of $\mathscr{L}$, '$[\![\phi]\!] = 1$ in V^Y' can be formalized. This is a consequence of Tarski's "undefinability of truth theorem", a theorem of logic which can safely be left to logicians: see [23, §3.5] and [44, p.125], for example. It accounts for the fact that certain results in this work are called "metatheorems".)

We now introduce notation for some specific formulae of $\mathscr{L}$. Let:

$\psi_\wedge(x_1,\ldots,x_4)$ (respectively, $\psi_\vee(x_1,\ldots,x_4)$) express

'x_4 is a Boolean algebra, x_1,x_2,x_3 belong to the underlying set of x_4, and $x_3 = x_1 \wedge x_2$ (respectively, $x_3 = x_1 \vee x_2$)';

$\psi_o(x_1,x_2)$ express

'x_2 is a complete Boolean algebra and x_1 belongs to V^{x_2}'.

$\psi_{\in}(x_1,\ldots,x_4)$ (respectively, $\psi_{=}(x_1,\ldots,x_4)$) express

'x_4 is a complete Boolean algebra, x_3 belongs to the underlying set of x_4, x_1, x_2 belong to V^{x_4}, and $x_3 = [\![x_1 \hat{\in} x_2]\!]$ (respectively, $x_3 = [\![x_1 \hat{=} x_2]\!]$) in V^{x_4}'.

Finally, for each formula $\psi(x_1,\ldots,x_n)$ of $\mathscr{L}$, there is a formula $\psi^*(x_1,\ldots,x_n,x_{n+1},x_{n+2})$ which expresses

'x_{n+2} is a complete Boolean algebra, x_{n+1} belongs to the underlying set of x_{n+2}, $x_1,\ldots,x_n$ belong to $V^{x_{n+2}}$, and $x_{n+1} = [\![\psi[x_1,\ldots,x_n]]\!]$ in $V^{x_{n+2}}$'.

We now return to the construction of models of ZFC + ¬CH, given a model of ZFC.

Let $\mathfrak{M} = (M,E)$ be a model of ZFC, and let $\mathfrak{b}$ in M be such that

$$\mathfrak{M} \models \text{'}\mathfrak{b} \text{ is a Boolean algebra'}.$$

Then there exists $b \in M$ such that

$$\mathfrak{M} \models \text{'}b \text{ is the underlying set of } \mathfrak{b}\text{'}.$$

As in Definition 4.12, we set

$$\tilde{a} = \{c \in M : cEa\} \quad (a \in M).$$

There are naturally defined operations $\tilde{\wedge}$ and $\tilde{\vee}$ such that $(\tilde{b},\tilde{\wedge},\tilde{\vee},',0,1)$ is a Boolean algebra (where $'$, 0, and 1 are defined using $\tilde{\wedge}$ and $\tilde{\vee}$): for $a,c \in \tilde{b}$, set

$$a \,\tilde{\wedge}\, c = d \quad \text{if} \quad \mathfrak{M} \models \psi_{\wedge}[a,c,d,\mathfrak{b}],$$

$$a \,\tilde{\vee}\, c = d \quad \text{if} \quad \mathfrak{M} \models \psi_{\vee}[a,c,d,\mathfrak{b}].$$

Thus, for example, $a \tilde{\wedge} c = d$ if and only if $\mathfrak{M} \models$ '$a \wedge c = d$ in $\mathfrak{b}$'. We write $\mathfrak{b}_{\mathfrak{M}}$ for the Boolean algebra $\langle \tilde{b}, \tilde{\wedge}, \tilde{\vee}, ', 0, 1 \rangle$.

Thus the operations in our Boolean algebra $\mathfrak{b}_{\mathfrak{M}}$ are exactly those that the model $\mathfrak{M}$ "thinks" are the operations in its Boolean algebra $\mathfrak{b}$. The points of M which appear *internally* to $\mathfrak{M}$ to be Boolean algebras have a canonical *external* manifestation as Boolean algebras.

7.14 EXAMPLE

Choose $a_o \in M$ such that $\mathfrak{M} \models$ '$a_o \neq \emptyset$'. Then $\tilde{a}_o \neq \emptyset$. Let $\mathfrak{b} \in M$ be such that

$$\mathfrak{M} \models \text{'}\mathfrak{b} = (P(a_o), \cap, \cup, ', \emptyset, a_o)\text{'}.$$

Then $\tilde{b} = \{c \in M : \mathfrak{M} \models \text{'}c \subset a_o\text{'}\} = \{c \in M : \tilde{c} \subset \tilde{a}_o\}$. Let $X = \{\tilde{c} : c \in \tilde{b}\}$. Then $(X, \cap, \cup, ', \emptyset, \tilde{a}_o)$ is a Boolean algebra. The map

$$\pi : c \mapsto \tilde{c}, \quad (\tilde{b}, \tilde{\wedge}, \tilde{\vee}, ', 0, 1) \to (X, \cap, \cup, ', \emptyset, \tilde{a}_o)$$

is a Boolean homomorphism. Since $\mathfrak{M} \models$ Extensionality, Example 4.13(i), shows that π is a bijection, and so it is a Boolean isomorphism.

Suppose now that

$$\mathfrak{M} \models \text{'}\mathfrak{b} \text{ is a } \textit{complete} \text{ Boolean algebra'}.$$

Then it is not necessarily the case that $\mathfrak{b}_{\mathfrak{M}}$ is complete. For suppose that $\mathfrak{M}$ is a countable model of ZFC, and choose $\mathfrak{b} \in M$ so that $\mathfrak{M} \models$ '$\mathfrak{b}$ is an infinite Boolean algebra'. Then $\mathfrak{b}_{\mathfrak{M}}$ is a countable, infinite Boolean algebra, and it is easily seen that such a Boolean algebra cannot be complete. In fact, generally, $\mathfrak{b}$ is complete in $\mathfrak{M}$ if, for each $a \in M$, the set $\tilde{a} \cap \tilde{b}$ has a supremum in $\mathfrak{b}_{\mathfrak{M}}$, whereas $\mathfrak{b}_{\mathfrak{M}}$ is complete only if each subset of $\tilde{b}$ has a supremum in $\mathfrak{b}_{\mathfrak{M}}$.

Thus, assume that $\mathfrak{b}$ is a complete Boolean algebra in $\mathfrak{M}$, and let $\mathbb{B}$ be the completion of the Boolean algebra $\mathfrak{b}_{\mathfrak{M}}$ (see Theorem 2.14); we identify $\mathfrak{b}_{\mathfrak{M}}$ with a Boolean subalgebra of $\mathbb{B}$, and we write B for the underlying set of $\mathbb{B}$. Define

$$M^* = \{a \in M : \mathfrak{M} \models \psi_o[a,\mathfrak{b}]\}$$
$$= \{a \in M : \mathfrak{M} \models \text{'}a \text{ belongs to } V^{\mathfrak{b}}\text{'}\},$$

and define maps $E_{\mathbb{B}} : (a,c) \mapsto [\![a \mathbin{\hat{\in}} c]\!]$ and $\sim_{\mathbb{B}} : (a,c) \mapsto [\![a \mathbin{\hat{=}} c]\!]$ from $M^* \times M^*$ into B as follows:

$$E_{\mathbb{B}}(a,c) = d \quad \text{if} \quad \mathfrak{M} \models \psi_{\in}[a,c,d,\mathfrak{b}];$$

$$\sim_{\mathbb{B}}(a,c) = d \quad \text{if} \quad \mathfrak{M} \models \psi_{=}[a,c,d,\mathfrak{b}].$$

Thus, for example, $E_{\mathbb{B}}(a,c) = d$ if and only if $\mathfrak{M} \models \text{'}[\![a \mathbin{\hat{\in}} c]\!] = d \text{ in } V^{\mathfrak{b}}\text{'}$.

7.15 THEOREM

Let $\mathfrak{M}$ be a model of ZFC such that

$$\mathfrak{M} \models \text{'}\mathfrak{b} \text{ is a complete Boolean algebra'},$$

and let $\mathfrak{M}_{\mathbb{B}} = (M^*, E_{\mathbb{B}}, \sim_{\mathbb{B}})$, in the above notation. Then:

(i) $\mathfrak{M}_{\mathbb{B}}$ is a $\mathbb{B}$-valued model;

(ii) for each formula $\phi(x_1,\ldots,x_n)$ of $\mathcal{L}$ and each $a_1,\ldots,a_n \in M^*$, $[\![\phi[a_1,\ldots,a_n]]\!]^{\mathfrak{M}_{\mathbb{B}}} \in \tilde{b}$ and

$$[\![\phi[a_1,\ldots,a_n]]\!]^{\mathfrak{M}_{\mathbb{B}}} = c \quad \text{if and only if}$$

$$\mathfrak{M} \models \phi^*[a_1,\ldots,a_n,c,\mathfrak{b}];$$

(iii) $\mathfrak{M}_{\mathbb{B}}$ is full.

Proof

(i) Each condition of Definition 7.1, when interpreted by the rules that we have given, becomes a condition which, by the corresponding statement in 7.8 or 7.9, is satisfied.

(ii) This is proved by induction on the length of formulae. The steps not involving quantifiers are immediate. We prove that, if the result holds for a formula $\psi(x_1)$, then it holds for $\exists x_1 \psi$.

Since the result holds for ψ, for each $a \in M^*$ $[\![\psi[a]]\!]^{\mathfrak{M}_B} \in \tilde{b}$ and $[\![\psi[a]]\!]^{\mathfrak{M}_B} = c$ if and only if $\mathfrak{M} \models \psi^*[a,c,\mathfrak{b}]$. For each $s \in M$ with $\tilde{s} \subset \tilde{b}$, the supremum $\bigvee\tilde{s}$ exists in $\mathfrak{b}_{\mathfrak{M}}$, and it equals that element d of $\tilde{b}$ such that

$$\mathfrak{M} \models \text{'}\bigvee s = d \text{ in } \mathfrak{b}\text{'}. \tag{8}$$

Choose $s \in M$ with $\mathfrak{M} \models \text{'}s = \{[\![\psi[a]]\!] : a \in V^{\mathfrak{b}}\}\text{'}$, and let $d \in \tilde{b}$ be such that (8) holds. Then

$$\mathfrak{M} \models \text{'}[\![\exists x_1 \psi]\!] = d \text{ in } \mathfrak{b}\text{'},$$

and so $d = \bigvee\tilde{s}$ in $\mathfrak{b}_{\mathfrak{M}}$. By the inductive hypothesis

$$\tilde{s} = \{[\![\psi[a]]\!]^{\mathfrak{M}_B} : a \in M^*\},$$

and this shows that the result holds for the formula $\exists x_1 \psi$.

(iii) This is immediate from (ii) and 7.12(ii). ∎

The above theorem shows that internal computations within $\mathfrak{M}$ of the value of $[\![\phi[a_1,\ldots,a_n]]\!]$ in $\mathfrak{b}$ give the same value as the computation of $[\![\phi[a_1,\ldots,a_n]]\!]^{\mathfrak{M}_B}$, and so we have the following result.

7.16 PROPOSITION

Let ψ be a sentence.

(i) Suppose that $\mathfrak{N} \models \psi$ for each model $\mathfrak{N}$. If $\mathfrak{M} \models$ ZFC, then

$\mathfrak{M} \models$ 'for each complete Boolean algebra $\mathbb{B}$, $[\![\psi]\!] = 1$ in $V^{\mathbb{B}}$'.

(ii) Suppose that $\mathfrak{N} \models \psi$ for each model $\mathfrak{N}$ such that $\mathfrak{N} \models$ ZFC. If $\mathfrak{M} \models$ ZFC, then

$\mathfrak{M} \models$ 'for each complete Boolean algebra $\mathbb{B}$, $[\![\psi]\!] = 1$ in $V^{\mathbb{B}}$'.

<u>Proof</u>

(i) Suppose that the result fails. Then there exists $\mathfrak{b}$ in M such that

$\mathfrak{M} \models$ '$[\![\psi]\!] \neq 1$ in $V^{\mathfrak{b}}$'.

Let $\mathfrak{M}_{\mathbb{B}}$ be obtained from $\mathfrak{M}$ and $\mathfrak{b}$ as above. Then $\mathfrak{M}_{\mathbb{B}}$ is a full $\mathbb{B}$-valued model, and $[\![\psi]\!] \neq 1$ in $\mathfrak{M}_{\mathbb{B}}$. Hence there is an ultrafilter U in $\mathbb{B}$ with $[\![\neg\psi]\!] \in U$. By 7.4, $\mathfrak{M}_{\mathbb{B}}(U) \models \neg\psi$. But this contradicts the hypothesis that $\mathfrak{N} \models \psi$ for each model $\mathfrak{N}$.

(ii) This is the same argument, save that it uses 7.13 to establish that $[\![\phi]\!] = 1$ in $\mathfrak{M}_{\mathbb{B}}$ for each axiom of ZFC. ∎

By using Gödel's completeness theorem, the above result can be reformulated as a metatheorem. Result (i) says that, if ψ is logically derivable, then it can be proved that $[\![\psi]\!] = 1$ in $V^{\mathbb{B}}$, and it justifies the fact that, in 7.13, it was sufficient to calculate the $\mathbb{B}$-value of an axiom equivalent to the given one.

Proposition 7.16(ii) shows that, if ZFC proves a sentence ϕ, then ZFC proves that $[\![\phi]\!] = 1$ in $V^{\mathbb{B}}$. Thus we may *reason within* $V^{\mathbb{B}}$ in the same way as we do in V: any normal mathematical argument can be carried out "within $[\![\ldots]\!]$". This is very important because it is exactly what we shall do at numerous points in the remainder of this book. Here is a first example: let $\mathbb{B}$ be a complete Boolean algebra, and take $u, v \in V^{\mathbb{B}}$ such that $[\![$ 'u is a ccc Boolean algebra' $]\!] = 1$ and such that $[\![$ 'v is an antichain in u' $]\!] = 1$. Then $[\![$ 'v is countable' $]\!] = 1$.

We now come to the key paragraph of this Chapter.

Let $\mathfrak{M}$ be a model of ZFC, and let ψ be a sentence which we wish to prove is consistent with ZFC. (We shall shortly take ψ to be $\neg$CH.) Suppose that we can prove that there is a complete Boolean algebra $\mathbb{B}$ such that $[\![\psi]\!] \neq 0$ in $V^{\mathbb{B}}$. Then

$$\mathfrak{M} \models \text{'there is a complete Boolean algebra } \mathbb{B} \text{ with } [\![\psi]\!] \neq 0 \text{ in } V^{\mathbb{B}}\text{'},$$

and so there exists $\mathfrak{b} \in M$ such that

$$\mathfrak{M} \models \text{'}\mathfrak{b} \text{ is a complete Boolean algebra and } [\![\psi]\!] \neq 0 \text{ in } V^{\mathfrak{b}}\text{'}.$$

Let $\mathfrak{b}_{\mathfrak{M}}$ be the Boolean algebra obtained from $\mathfrak{b}$, as above, let $\mathbb{B}$ be the completion of $\mathfrak{b}_{\mathfrak{M}}$, and let $\mathfrak{M}_{\mathbb{B}} = (M^*, E_{\mathbb{B}}, \sim_{\mathbb{B}})$ be the corresponding $\mathbb{B}$-valued model. By 7.15, $\mathfrak{M}_{\mathbb{B}}$ is full and

$$[\![\psi]\!]^{\mathfrak{M}_{\mathbb{B}}} \neq 0.$$

Thus there is an ultrafilter $\mathcal{U}$ in $\mathbb{B}$ such that $[\![\psi]\!]^{\mathfrak{M}_{\mathbb{B}}} \in \mathcal{U}$. By 7.4,

$$\mathfrak{M}_{\mathbb{B}}(\mathcal{U}) \models \text{ZFC} + \psi.$$

We shall see shortly, on page 164, that $\mathfrak{M}_{\mathbb{B}}(\mathcal{U})$ is an extension of $\mathfrak{M}$. Thus $\mathfrak{M}_{\mathbb{B}}(\mathcal{U})$ is the model that we have been seeking that witnesses the consistency of ZFC + ψ.

Our entire development of logic, including the theory of models and of Boolean-valued models, was to justify the following claim.

Let ψ *be a sentence of* $\mathcal{L}$. *Suppose that it can be proved (in* ZFC*) that 'there is a complete Boolean algebra* B *with* $[\![\psi]\!] \neq 0$ *in* V^{B}*'. Then* Con ZFC *implies* Con(ZFC + ψ).

A forcing argument is essentially a construction of a complete Boolean algebra $\mathcal{B}$, together with a computation that $[\![\psi]\!] \neq 0$ in $V^{\mathcal{B}}$. The process will be similar to a normal proof in set theory, but there is a further freedom given by the possibility of choosing $\mathcal{B}$. It is possible to learn and to carry out the technicalities of a forcing argument whilst remaining ignorant of the justification that the argument gives the required independence result.

Let $\mathcal{B}$ be a complete Boolean algebra. Since we now know that $[\![\phi]\!] = 1$ for each axiom ϕ of ZFC, we can obtain manifestations within $V^{\mathcal{B}}$ of familiar sets constructed in the theory ZFC - for example, of ω, of ω_1, and of $\mathbb{R}$. This leads to a study of ordinals and cardinals in $V^{\mathcal{B}}$, and in particular to the construction of a Boolean algebra $\mathcal{B}$ such that $[\![\neg\mathrm{CH}]\!] = 1$. We shall build a class $\underset{\sim}{\mathrm{Ord}}^{\vee}$ in $V^{\mathcal{B}}$ from the class $\underset{\sim}{\mathrm{Ord}}$ of ordinals such that, if

$$[\![\text{'}u \text{ is an ordinal'}]\!] \neq 0,$$

then $[\![u \hat{=} v]\!] \neq 0$ for some $v \in \underset{\sim}{\mathrm{Ord}}^{\vee}$. To do this, we shall represent V inside $V^{\mathcal{B}}$.

7.17 DEFINITION

Let $\mathcal{B}$ be a complete Boolean algebra. A map $^{\vee} : V \to V^{\mathcal{B}}$ is defined by recursion on the rank of elements of V. Set $\emptyset^{\vee} = \emptyset$, and set

$$\mathrm{dom}\, \check{u} = \{\check{v} : v \in u\}, \quad \check{u}(t) = 1 \quad (t \in \mathrm{dom}\, \check{u}).$$

The map $^\vee$ exists and is uniquely specified by the conditions. Note that, for $u \in V^{B}$ and $v \in V$,

$$[\![u \mathbin{\hat{\in}} \check{v}]\!] = \bigvee\{[\![u \mathbin{\hat{=}} \check{w}]\!] : w \in v\}. \tag{9}$$

Recall that rk(u) denotes the rank of an element u of V, as on page 85.

7.18 LEMMA

Let $u, v \in V$. Then:

(i) $u \in v$ if and only if $[\![\check{u} \mathbin{\hat{\in}} \check{v}]\!] = 1$,

$u \notin v$ if and only if $[\![\check{u} \mathbin{\hat{\in}} \check{v}]\!] = 0$;

(ii) $u = v$ if and only if $[\![\check{u} \mathbin{\hat{=}} \check{v}]\!] = 1$,

$u \neq v$ if and only if $[\![\check{u} \mathbin{\hat{=}} \check{v}]\!] = 0$.

Proof

We prove the two statements simultaneously by induction on the position of $(\mathrm{rk}(u), \mathrm{rk}(v))$ in $(\mathbf{Ord}^2, <)$.

The statements are trivial if $u = \emptyset$ or if $v = \emptyset$.

Now suppose that the statements hold for $r, s \in V$ with $(\mathrm{rk}(r), \mathrm{rk}(s)) < (\mathrm{rk}(u), \mathrm{rk}(v))$. Then, by (9),

$$[\![\check{u} \mathbin{\hat{\in}} \check{v}]\!] = \bigvee\{[\![\check{u} \mathbin{\hat{=}} \check{w}]\!] : w \in v\}$$
$$= \begin{cases} 1 & \text{if } u \in v, \\ 0 & \text{if } u \notin v. \end{cases}$$

Thus (i) follows. The proof of (ii) is similar, noting that, if $w \in \mathrm{dom}\, u$, then $(\mathrm{rk}\, v, \mathrm{rk}\, w) < (\mathrm{rk}\, u, \mathrm{rk}\, v)$ in $(\mathbf{Ord}^2, <)$, and so (i) holds for v and w. ∎

Suppose that $\mathfrak{M} = (M, E) \models \mathrm{ZFC}$, and let $\mathfrak{b} \in M$ with

$\mathfrak{M} \models$ '$\mathfrak{b}$ is a complete Boolean algebra'.

Take $\mathfrak{M}_B = (M^*, E_B, \sim_B)$ to be as above, and define $\pi : M \to M^*$ by

$$\pi(a) = r \quad \text{if} \quad \mathfrak{M} \models \psi[\mathfrak{b}, r, a],$$

where $\psi(x,y,z)$ expresses 'x is a complete Boolean algebra, $y \in V^x$, and $y = \check{z}$'. Take $a,c \in M$ with aEc. Then, by 7.18,

$$[\![\pi(a) \ \hat{\in} \ \pi(c)]\!]^{\mathfrak{M}_B} = 1.$$

This, and the obvious three related formulae, show that π induces an isomorphism from $\mathfrak{M}$ onto a submodel of $\mathfrak{M}_B(U)$, and so $\mathfrak{M}_B(U)$ is indeed an extension of $\mathfrak{M}$.

7.19 LEMMA

Let $u,v \in V$. Then $[\![\check{u} \subset \check{v}]\!] = 1$ (respectively, 0) if and only if $u \subset v$ (respectively, $u \not\subset v$).

<u>Proof</u>

This is immediate from 7.18, on expanding $[\![\check{u} \subset \check{v}]\!]$ by formula (3) of page 139. ∎

7.20 LEMMA

Let $u \in V$. Then:

(i) u is transitive if and only if

$$[\!['\check{u} \text{ is transitive}']\!] = 1,$$

u is not transitive if and only if

$$[\!['\check{u} \text{ is transitive}']\!] = 0;$$

(ii) $u \in \underset{\sim}{\mathrm{Ord}}$ if and only if

$$[\!['\check{u} \text{ is an ordinal}']\!] = 1,$$

$u \notin \underset{\sim}{\mathrm{Ord}}$ if and only if

$$[\!['\check{u} \text{ is an ordinal}']\!] = 0.$$

<u>Proof</u>

(i) Suppose that u is transitive, and take

$v \in V^{B}$. For $w \in u$, we have $w \subset u$, and so, by 7.19, $[\![\check{w} \subset \check{u}]\!] = 1$. Thus, using (9) and 7.12(i),

$$[\![v \,\hat{\in}\, \check{u}]\!] = \bigvee\{[\![v \,\hat{=}\, \check{w}]\!] \wedge [\![\check{w} \subset \check{u}]\!] : w \in u\}$$

$$\leqslant [\![v \subset \check{u}]\!],$$

and so $[\![\text{'}\check{u} \text{ is transitive'}]\!] = 1$.

Similarly, if u is not transitive, then we have $[\![\text{'}\check{u} \text{ is transitive'}]\!] = 0$.

(ii) Suppose that $u \in \underset{\sim}{\mathrm{Ord}}$. By (i), to show that $[\![\text{'}\check{u} \text{ is an ordinal'}]\!] = 1$, it suffices to show that, for $v, w \in V^{B}$,

$$[\![v \in \check{u}]\!] \wedge [\![w \in \check{u}]\!]$$
$$\leqslant [\![v \,\hat{\in}\, w]\!] \vee [\![v \,\hat{=}\, w]\!] \vee [\![w \,\hat{\in}\, v]\!]. \tag{10}$$

If $v, w \in \operatorname{dom} \check{u}$, then (10) holds because $v = \check{r}$ and $w = \check{s}$ for some $r, s \in u$. The general case then follows from (9).

The case where $u \notin \underset{\sim}{\mathrm{Ord}}$ is again similar. ∎

7.21 LEMMA

Let $\alpha \in \underset{\sim}{\mathrm{Ord}}$, and let $u \in V^{B}$.

(i) If $\alpha \geqslant \rho(u)$, then $[\![\check{\alpha} \,\hat{\in}\, u]\!] = 0$.

(ii) If $\alpha > \rho(u)$, then $[\![\check{\alpha} \,\hat{=}\, u]\!] = 0$.

Proof

We prove (i) and (ii) simultaneously by induction on the position of $(\alpha, \rho(u))$ in $(\underset{\sim}{\mathrm{Ord}}^{2}, <)$.

The statements are trivial if $\alpha = \emptyset$ or if $u = \emptyset$. Now suppose that they hold for β, v with $(\beta, \rho(v)) < (\alpha, \rho(u))$. If $\alpha \geqslant \rho(u)$, then $\rho(t) < \alpha \quad (t \in \operatorname{dom} u)$, and so $[\![\check{\alpha} \,\hat{=}\, t]\!] = 0$, whence $[\![\check{\alpha} \,\hat{\in}\, u]\!] = 0$. If $\alpha > \rho(u)$, set $\beta = \rho(u)$. Then

$$[\![\check{\alpha} \mathrel{\hat{=}} u]\!] \leqslant [\![\check{\alpha} \subset u]\!] = \bigwedge\{[\![\check{w} \mathrel{\hat{\in}} u]\!] : w \in \alpha\}$$

$$\leqslant [\![\beta \mathrel{\hat{\in}} u]\!] = 0.$$

The induction continues. ∎

7.22 LEMMA

Let $u \in V^{\mathbf{B}}$, and suppose that

$$[\![\text{'}u \text{ is an ordinal'}]\!] \neq 0.$$

Then there exists $\alpha \in \underset{\sim}{\mathrm{Ord}}$ with $[\![u \mathrel{\hat{=}} \check{\alpha}]\!] \neq 0$.

Proof

For $\beta \in \underset{\sim}{\mathrm{Ord}}$, $[\![\text{'}\check{\beta} \text{ is an ordinal'}]\!] = 1$ by 7.20(ii), and so

$$[\![u \mathrel{\hat{\in}} \check{\beta}]\!] \vee [\![u \mathrel{\hat{=}} \check{\beta}]\!] \vee [\![\check{\beta} \mathrel{\hat{\in}} u]\!]$$

$$\geqslant [\![\text{'}u \text{ is an ordinal'}]\!] \wedge [\![\text{'}\check{\beta} \text{ is an ordinal'}]\!]$$

$$= [\![\text{'}u \text{ is an ordinal'}]\!] \neq 0.$$

Take $\beta \in \underset{\sim}{\mathrm{Ord}}$ with $\beta > \rho(u)$. By 7.21, $[\![\check{\beta} \mathrel{\hat{=}} u]\!] = [\![\check{\beta} \mathrel{\hat{\in}} u]\!] = 0$, and so $[\![u \mathrel{\hat{\in}} \check{\beta}]\!] \neq 0$. By (9), there exists $\alpha < \beta$ with $[\![u \mathrel{\hat{=}} \check{\alpha}]\!] \neq 0$. ∎

7.23 DEFINITION

Set $\underset{\sim}{\mathrm{Ord}}^{\vee} = \{\check{\alpha} : \alpha \in \underset{\sim}{\mathrm{Ord}}\}$.

Then $\underset{\sim}{\mathrm{Ord}}^{\vee}$ is a class which is contained in $V^{\mathbf{B}}$, and 7.22 shows that it satisfies the condition specified on page 162: in fact,

$$[\![\text{'}u \text{ is an ordinal'}]\!] = \bigvee\{[\![u \mathrel{\hat{=}} \check{\alpha}]\!] : \alpha \in \underset{\sim}{\mathrm{Ord}}\}$$

$$= \bigvee\{[\![u \mathrel{\hat{=}} \check{\alpha}]\!] : \alpha \leqslant \rho(u)\}.$$

A variant of Lemma 7.22 looks plausible: "if

$[\![\text{'}u \text{ is an ordinal'}]\!] = 1$,

then there exists $\alpha \in \underset{\sim}{\text{Ord}}$ with $[\![u \mathrel{\hat{=}} \check{\alpha}]\!] = 1$". But this is false whenever $\mathcal{B} \neq (2)$. For take $b \in B\backslash\{0,1\}$. It follows from the mixing lemma that there exists $u \in V^{\mathcal{B}}$ with $[\![u \mathrel{\hat{=}} \check{2}]\!] \geqslant b$ and $[\![u \mathrel{\hat{=}} \check{3}]\!] \geqslant b'$, and then $[\![\text{'}u \text{ is an ordinal'}]\!] = 1$.

Many of the basic properties of sets are preserved in the passage from V to $V^{\mathcal{B}}$. For example, if u is an ordered pair, then

$[\![\text{'}\check{u} \text{ is an ordered pair'}]\!] = 1$:

to see this, one formalizes the statement 'u is an ordered pair' and uses (9) and 7.18. We list some more examples of properties that are preserved.

7.24 EXAMPLE

(i) If $u = v \times w$, then $[\![\text{'}\check{u} \mathrel{\hat{=}} \check{v} \times \check{w}\text{'}]\!] = 1$.

(ii) If $g: u \to v$ is a function, then

$[\![\text{'}\check{g} \text{ is a function from } \check{u} \text{ to } \check{v}\text{'}]\!] = 1$.

(iii) If $g: u \to v$ is an injection (respectively, a surjection), then

$[\![\text{'}\check{g} \text{ is an injection (respectively, a surjection), from } \check{u} \text{ to } \check{v}\text{'}]\!] = 1$.

(iv) $[\![\text{'}(\alpha+1)^{\vee} \text{ is the minimum ordinal } > \check{\alpha}\text{'}]\!] = 1$.

(v) $[\![\text{'}\check{\omega} \text{ is the least infinite ordinal'}]\!] = 1$.

<u>Proof</u>

It is easy to check Examples (i) - (iii). If (iv) were false, there would exist β with

$$[\![\check{\alpha} \,\hat{\in}\, \check{\beta}]\!] \wedge [\![\check{\beta} \,\hat{\in}\, (\alpha+1)^{\vee}]\!] \neq 0,$$

and, by 7.18, this is impossible. Example (v) follows from (iv). ∎

However, other basic properties of sets may not be preserved in the passage from V to V^{B}. These include the power set operation (in general, $[\![\,'P(\check{u}) = (P(u))^{\vee}\,']\!] \neq 1$) and the cardinality of a set. There is a general theorem that asserts that properties specified by "Δ_o-formulae" are preserved: we shall return to this point in a more general context in Chapter 8.

It follows from 7.24(iii) and (v) that, if α is a countable ordinal, then $[\![\,'\check{\alpha}$ is countable$'\,]\!] = 1$. However, if $\check{\alpha}$ is an uncountable ordinal, it is not always true that $[\![\,'\check{\alpha}$ is uncountable$'\,]\!] \neq 0$: it is particularly important for us to deal with this possibility. More generally, if $u,v \in V$, then

$$|u| = |v| \text{ implies that } [\![\,'|\check{u}| = |\check{v}|'\,]\!] = 1,$$

but

$$|u| \neq |v| \text{ does not imply that } [\![\,'|\check{u}| \neq |\check{v}|'\,]\!] = 1.$$

7.25 DEFINITION

Let κ be a cardinal. Then κ is <u>preserved in V^{B}</u> if $[\![\,'\check{\kappa}$ is a cardinal$'\,]\!] = 1$.

We have seen that $\aleph_o$ is preserved in V^{B}. If $\aleph_1$ is preserved, then

$$[\![\,'\check{\aleph}_1 \text{ is the least uncountable cardinal}'\,]\!] = 1,$$

and we write this as $[\![\,'\check{\aleph}_1 = \aleph_1'\,]\!] = 1$. (To recap, in this expression $\check{\aleph}_1$ is the image under the embedding $^\vee$ of the first uncountable ordinal in V, whereas $\aleph_1$ is the first uncountable ordinal in the universe $V^{\mathbf{B}}$.) If $\aleph_2$ is preserved, then all we can say in general is that

$$[\![\text{'}\check{\aleph}_2 = \aleph_2 \text{ or } \check{\aleph}_2 = \aleph_1\text{'}]\!] = 1,$$

but, if $\aleph_1$ and $\aleph_2$ are preserved, then $[\![\text{'}\check{\aleph}_2 = \aleph_2\text{'}]\!] = 1$.

To study the preservation of cardinals, we consider terms for subsets of $\check{u}$ and for functions from $\check{u}$ to $\check{v}$, where $u,v \in V$.

Let $u \in V$. For $v \in V^{\mathbf{B}}$, we set

$$f_v : t \mapsto [\![\check{t} \mathbin{\hat{\in}} v]\!], \quad u \to B,$$

and, for $f\colon u \to B$, we define $v_f \in V^{\mathbf{B}}$ by

$$\operatorname{dom} v_f = \{\check{s} : s \in u\}, \quad v_f(\check{s}) = f(s).$$

Then $[\![\check{s} \mathbin{\hat{\in}} v_f]\!] = f(s)$ and $[\![v_f \subset \check{u}]\!] = 1$. For $v \in V^{\mathbf{B}}$, set $g = f_v$ and $v^* = v_g$. Then $[\![v \mathbin{\hat{=}} v^*]\!] = [\![v \subset \check{u}]\!]$, and so $v \sim v^*$ if $[\![v \subset \check{u}]\!] = 1$. Thus, up to equivalence, we have obtained a canonical set of terms for subsets of $\check{u}$.

We essentially repeat this construction in the next theorem, obtaining a canonical set of terms for bijections.

7.26 THEOREM

Let $u,v \in V$. Then $[\![\text{'}|\check{u}| = |\check{v}|\text{'}]\!] \geqslant b$ in $V^{\mathbf{B}}$ if and only if there is a function $f\colon u \times v \to B$ such that:

(i) $\bigvee\{f(r,s) : s \in v\} = b \quad (r \in u)$;

(ii) $\bigvee\{f(r,s) : r \in u\} = b \quad (s \in v)$;

(iii) $f(r,s_1) \wedge f(r,s_2) = 0 \quad (r \in u,\ s_1 \neq s_2 \text{ in } v)$;

(iv) $f(r_1,s) \wedge f(r_2,s) = 0 \quad (s \in v,\ r_1 \neq r_2 \text{ in } u)$.

<u>Proof</u>

Suppose that $[\!['|\check{u}| = |\check{v}|']\!] \geq b$. By the fullness lemma, there exists $w \in V^{\mathbf{B}}$ with

$$[\!['w \text{ is a bijection from } \check{u} \text{ to } \check{v}']\!] \geq b.$$

Define $f: u \times v \to B$ by setting

$$f(r,s) = [\!['\check{s} \text{ is the value of } w \text{ at } \check{r}']\!] \wedge b.$$

Then f satisfies conditions (i) - (iv): (i) holds because $[\!['\text{dom } w = \check{u}']\!] \geq b$, (ii) holds because $[\!['\text{ran } w = \check{v}']\!] \geq b$, (iii) holds because $[\!['w \text{ is a function}']\!] \geq b$, and (iv) holds because $[\!['w \text{ is an injection}']\!] \geq b$.

Conversely, suppose that $f: u \times v \to B$ satisfies (i) - (iv). Define $w \in V^{\mathbf{B}}$ by

$$\text{dom } w = \{(r,s)^{\vee} : r \in u,\ s \in v\}, \quad w((r,s)^{\vee}) = f(r,s).$$

Clearly $[\!['w \subset \check{u} \times \check{v}']\!] = 1$, and it is easily checked that

$$[\!['w \text{ is a bijection from } \check{u} \text{ to } \check{v}']\!] = b.$$

Thus $[\!['|\check{u}| = |\check{v}|']\!] \geq b$, as required. ∎

We can now show that cardinals are preserved in $V^{\mathbf{B}}$ *provided that* $\mathbf{B}$ satisfies the key countable chain condition introduced in Chapter 4. Recall that $\mathbf{B}$ is ccc if each antichain in B is countable.

7.27 THEOREM

Let $\mathbf{B}$ be a complete ccc Boolean algebra. Then each cardinal is preserved in $V^{\mathbf{B}}$.

<u>Proof</u>

Let κ be a cardinal, and let σ be an ordinal with $\sigma < \kappa$. Set $b = [\!['|\check{\sigma}| = |\check{\kappa}|']\!]$. If $\kappa = \aleph_0$, then

$b = 0$ by 7.24(v).

Now suppose that $\kappa > \aleph_0$. Let $f: \sigma \times \kappa \to B$ be the function specified in 7.26. For $\alpha < \sigma$, set $f_\alpha(\beta) = f(\alpha, \beta)$ $(\beta < \kappa)$, and set

$$S_\alpha = \{\beta : f(\alpha,\beta) \neq 0\}, \quad S = \bigcup S_\alpha.$$

By (iii) in 7.26, $f_\alpha : S_\alpha \to B\backslash\{0\}$ is an injection and $f_\alpha(S_\alpha)$ is an antichain in B. Since $\mathbb{B}$ is ccc, S_α is countable, and so $|S| \leqslant \max\{\sigma, \aleph_0\} < \kappa$. Take $\beta < \kappa$ such that $\beta \notin S$. Then $b \leqslant \bigvee\{f(\alpha,\beta) : \alpha < \sigma\} = 0$.

Thus, in each case $[\![\text{'}|\check{\sigma}| = |\check{\kappa}|\text{'}]\!] = 0$, and so $[\![\text{'}\check{\kappa} \text{ is a cardinal'}]\!] = 1$. This shows that κ is preserved in $V^{\mathbb{B}}$. ∎

Let $\mathbb{B}$ be any complete Boolean algebra. The above proof shows that, if κ is a cardinal with $\kappa > |B|$, then κ is preserved in $V^{\mathbb{B}}$.

7.28 THEOREM

Let $\mathbb{B}$ be a complete Boolean algebra, and set $\kappa = |B|^{\aleph_0}$. Then

$$[\![\text{'}2^{\aleph_0} \leqslant \check{\kappa}\text{'}]\!] = 1 \text{ in } V^{\mathbb{B}}.$$

<u>Proof</u>

Set $\lambda = \kappa^+$, the successor cardinal to κ. We first *claim* that $[\![\text{'}2^{\aleph_0} \geqslant \check{\lambda}\text{'}]\!] = 0$ in $V^{\mathbb{B}}$. If this is not the case, there exists $b \in B\backslash\{0\}$ and $t \in V^{\mathbb{B}}$ such that

$$[\![\text{'}t \text{ is an injection from } \check{\lambda} \text{ into } P(\check{\mathbb{N}})\text{'}]\!] \geqslant b.$$

By the fullness lemma, for each $\alpha < \lambda$, there exists $v_\alpha \in V^{\mathbb{B}}$ such that

$$[\![\text{'the value of } t \text{ at } \check{\alpha} \text{ is } v_\alpha\text{'}]\!] \geqslant b,$$

and then $[\!['v_\alpha \subset \check{\mathbb{N}}']\!] \geq b$ $(\alpha < \lambda)$. For $\alpha < \lambda$, set

$$f_\alpha : n \mapsto [\![\check{n} \hat{\in} v_\alpha]\!], \quad \mathbb{N} \to B.$$

If $f_\alpha = f_\beta$, then $[\![v_\alpha \hat{=} v_\beta]\!] \geq b$, and so $\alpha = \beta$ because $[\!['t$ is an injection$']\!] \geq b$. Thus $H : \alpha \mapsto f_\alpha$, $\lambda \to B^{\mathbb{N}}$, is an injection, and so $\lambda \leq |B^{\mathbb{N}}| = |B|^{\aleph_0}$, a contradiction of the hypothesis that $\kappa \leq |B|^{\aleph_0}$. Hence the claim holds.

The result is immediate from the claim. ∎

It can be shown that, if $\mathcal{B}$ is a complete Boolean algebra and $|B|$ is infinite, then $|B|^{\aleph_0} = |B|$.

It follows from 7.26 that, if $\kappa \geq |B|$, then κ is preserved in $V^{\mathcal{B}}$, and it then follows easily from 7.28 that, if $|B| = 2^{\aleph_0}$, then $[\!['2^{\aleph_0} = (2^{\aleph_0})^{\vee}']\!] = 1$ in $V^{\mathcal{B}}$. We shall not need this remark, but we shall need the following corollary of the theorem.

7.29 COROLLARY (CH)

Let $\mathcal{B}$ be a complete Boolean algebra.

(i) If $|B| \leq \aleph_2$, then $[\!['2^{\aleph_0} \leq \aleph_2']\!] = 1$ in $V^{\mathcal{B}}$.

(ii) If $|B| \leq \aleph_1$, then $[\![\mathrm{CH}]\!] = 1$ in $V^{\mathcal{B}}$.

Proof

(i) Since CH holds, $\aleph_2^{\aleph_0} = \aleph_2$ by 5.11, and so $|B|^{\aleph_0} \leq \aleph_2^{\aleph_0} = \aleph_2$. By the theorem, $[\!['2^{\aleph_0} \leq \check{\aleph}_2']\!] = 1$. Certainly $[\!['\check{\aleph}_2 \leq \aleph_2']\!] = 1$, and so the result follows.

(ii) Again using CH, $|B|^{\aleph_0} = \aleph_1$, and so $[\!['2^{\aleph_0} \leq \check{\aleph}_1']\!] = 1$. Certainly $[\!['\check{\aleph}_1 \leq \aleph_1']\!] = 1$, and so $[\!['2^{\aleph_0} \leq \aleph_1']\!] = 1$, i.e., $[\![\mathrm{CH}]\!] = 1$. ∎

We can now construct the complete Boolean algebra

$\mathcal{B}$ such that $[\![\mathrm{CH}]\!] = 0$ in $V^{\mathcal{B}}$. We must show that

$$[\![\text{'}|P(\mathbb{N}^{\vee})| > \aleph_1\text{'}]\!] = 1 \quad \text{in} \quad V^{\mathcal{B}}.$$

The following theorem gives a condition which reduces the problem to a combinatorial question about Boolean algebras.

Let $\mathcal{B}$ be a complete Boolean algebra, let $f,g : \mathbb{N} \to B$ be functions, and let v_f, v_g be the corresponding terms, (see page 169), so that $\mathrm{dom}\, v_f = \mathrm{dom}\, v_g = \{\check{n} : n \in \mathbb{N}\}$, $v_f(\check{n}) = f(n)$, and $v_g(\check{n}) = g(n)$. Then

$$[\![v_f \subset v_g]\!] = \bigwedge\{v_f(\check{n})' \vee [\![\check{n} \mathbin{\hat{\in}} v_g]\!] : n \in \mathbb{N}\}$$

$$= \bigwedge\{f(n)' \vee g(n) : n \in \mathbb{N}\},$$

and so

$$[\![v_f \mathbin{\hat{\neq}} v_g]\!] = \bigvee\{f(n) \vartriangle g(n) : n \in \mathbb{N}\}, \qquad (11)$$

where $a \vartriangle b = (a' \wedge b) \vee (a \wedge b')$ $(a,b \in B)$. This shows that condition (12), below, arises naturally.

7.30 THEOREM

Let $\mathcal{B}$ be a complete ccc Boolean algebra. Suppose that there is a set $S \subset B^{\mathbb{N}}$ such that $|S| = \aleph_2$ and such that

$$\bigvee\{f(n) \vartriangle g(n) : n \in \mathbb{N}\} = 1 \quad (f,g \in S,\ f \neq g). \quad (12)$$

Then $[\![\mathrm{CH}]\!] = 0$ in $V^{\mathcal{B}}$.

<u>Proof</u>

Suppose, if possible, that $[\![\mathrm{CH}]\!] = b \neq 0$ in $V^{\mathcal{B}}$. By 7.28, $[\![\text{'}\check{\aleph}_1 = \aleph_1\text{'}]\!] = 1$, and so

$$[\![\text{'there is a surjection from } \check{\aleph}_1 \text{ onto } P(\mathbb{N}^{\vee})\text{'}]\!] = b.$$

By the fullness lemma, there exists $w \in V^{\mathcal{B}}$ with

$$[\![\text{'}w \text{ is a surjection from } \check{\aleph}_1 \text{ onto } P(\mathbb{N}^\vee)\text{'}]\!] = b.$$

Since $[\![\text{'}v_f \in P(\mathbb{N}^\vee)\text{'}]\!] = [\![v_f \subset \mathbb{N}^\vee]\!] = 1$ for $f \in S$, there exists $\alpha_f < \omega_1$ such that

$$[\![\text{'}v_f \text{ is the value of } w \text{ at } \check{\alpha}_f\text{'}]\!] \neq 0.$$

Since $|S| = \aleph_2$, there is a subset T of S with $|T| = \aleph_2$ and $\alpha < \omega_1$ such that $b_f \neq 0$, where, for $f \in T$,

$$b_f = [\![\text{'}v_f \text{ is the value of } w \text{ at } \check{\alpha}\text{'}]\!].$$

Take $f,g \in T$ with $f \neq g$. By (11) and (12), $[\![v_f \hat{=} v_g]\!] = 0$, and so $b_f \wedge b_g = 0$. Thus $\{b_f : f \in T\}$ is an antichain in $\mathcal{B}$ of cardinality $\aleph_2$. But this is a contradiction of the hypothesis that $\mathcal{B}$ is ccc, and so $b = 0$, as required. ∎

Our choice of a Boolean algebra $\mathcal{B}$ to satisfy the conditions of Theorem 7.30 was implicitly given in the original arguments of Cohen.

7.31 DEFINITION

Let $2^{\mathbb{N}}$ be the Cantor set $\{0,1\}^{\mathbb{N}}$, and set $X_C = (2^{\mathbb{N}})^{\omega_2}$.

The set X_C is a compact Hausdorff space with respect to the product topology. Of course, X_C is homeomorphic to 2^{ω_2}, the Cantor cube of weight ω_2. We regard X_C as the set of functions $F: \omega_2 \times \mathbb{N} \to \{0,1\}$; for $\alpha < \omega_2$, set $F_\alpha(n) = F(\alpha,n)$.

The regular-open algebra $R(X)$ of a topological space X was introduced in Example 2.7, and it was noted there that $R(X)$ is a complete Boolean algebra.

7.32 PROPOSITION

The Boolean algebra $R(X_C)$ is ccc.

<u>Proof</u>

Let τ be the standard base for the topology of $2^{\mathbb{N}}$. Clearly τ is countable.

Let P be the collection of maps whose domain is a finite subset of ω_2 and whose range is contained in τ. For $t \in P$, set

$$U_t = \{F \in X_C : F_\alpha \in t(\alpha) \quad (\alpha \in \text{dom } t)\}.$$

Then U_t is a clopen set in X_C, and $\{U_t : t \in P\}$ is a base for the topology of X_C. Note that, if $s,t \in P$ and if $s(\alpha) = t(\alpha)$ $(\alpha \in \text{dom } s \cap \text{dom } t)$, then $U_s \cap U_t \neq \emptyset$.

Suppose, if possible, that $R(X_C)$ is not ccc. Then there exists an uncountable subset A of P such that $U_s \cap U_t = \emptyset$ whenever $s,t \in A$ with $s \neq t$. Let $F = \{\text{dom } t : t \in A\}$. Suppose that F is uncountable. Then, by the Δ-system lemma, Proposition 6.1, there is a finite subset T of ω_2 and an uncountable subfamily G of F such that $\text{dom } s \cap \text{dom } t = T$ whenever $\text{dom } s$ and $\text{dom } t$ are distinct elements of F. Thus there is an uncountable subset A^* of A such that $\text{dom } s \cap \text{dom } t = T$ whenever s and t are distinct elements of A^*. This conclusion also holds if F is countable.

Since τ is countable, we may suppose that, for $s,t \in A^*$ and $\alpha \in T$, $s(\alpha) = t(\alpha)$. But this implies that $U_s \cap U_t \neq \emptyset$ $(s,t \in A^*)$, a contradiction of the hypothesis. Thus $R(X_C)$ is ccc. ∎

7.33 THEOREM

There is a complete Boolean algebra $\mathcal{B}$ such that $[\![CH]\!] = 0$ in $V^{\mathcal{B}}$.

<u>Proof</u>

Set $\mathcal{B} = R(X_C)$, so that $\mathcal{B}$ is a complete ccc

Boolean algebra. We shall apply Theorem 7.30.

For $\alpha < \omega_2$, define $f_\alpha : \mathbb{N} \to P(X_C)$ by setting

$$f_\alpha(n) = \{F \in X_C : F(\alpha,n) = 1\} \quad (n \in \mathbb{N}).$$

Clearly each $f_\alpha(n)$ is clopen in X_C, and so $f_\alpha(n) \in B$. Set $S = \{f_\alpha : \alpha < \omega_2\}$. Since $f_\alpha \neq f_\beta$ if $\alpha \neq \beta$, $|S| = \aleph_2$.

Take $\alpha, \beta < \omega_2$ with $\alpha \neq \beta$. Then

$$\begin{aligned} &\bigcup\{f_\alpha(n) \vartriangle f_\beta(n) : n \in \mathbb{N}\} \\ &\quad = \{F \in X_C : F(\alpha,n) \neq F(\beta,n) \quad \text{for some} \quad n \in \mathbb{N}\}, \end{aligned}$$

a set which is clearly dense in X_C. It follows from the first of equations (3) of Chapter 2 (page 29) that

$$\bigvee\{f_\alpha(n) \vartriangle f_\beta(n) : n \in \mathbb{N}\} = 1,$$

and so S satisfies the conditions in 7.30. The result follows. ∎

We have now achieved the following goal.

7.34 THEOREM

Assume that there is a model $\mathfrak{M}$ of ZFC. Then there is a model, extending $\mathfrak{M}$, of ZFC + ¬CH. ∎

We noted in Theorem 4.19 the result of Gödel that, if there is a model $\mathfrak{M}$ of ZFC, then there is a submodel of $\mathfrak{M}$ which is a model of ZFC + CH. We shall prove, in Theorem 7.38, that, if there is a model of ZFC then there is a model of ZFC + CH; we shall also prove a related result, Theorem 7.39, that will be required in Chapter 8.

7.35 DEFINITION

Let κ be a cardinal. Then $Q(\kappa)$ is the set of injective maps f such that $\operatorname{dom} f \in \kappa^+$ and such that $\operatorname{ran} f \subset P(\kappa)$, the power set of κ. For $f, g \in Q(\kappa)$, set

$f \leq g$ if $\operatorname{dom} g \subset \operatorname{dom} f$ and $f|\operatorname{dom} g = g$.

Certainly $(Q(\kappa),\leq)$ is a partially ordered set. By 2.10, $Q(\kappa)$ has a completion: it is denoted by $(\mathcal{B}(\kappa),\pi)$. The partially ordered set $(Q(\kappa),\leq)$ is clearly separative (see page 31), and so π is an injection; we identify $Q(\kappa)$ with its image in $B(\kappa)$, so that $Q(\kappa)$ is dense in $B(\kappa)\setminus\{0\}$.

7.36 **LEMMA**

Let $\mathcal{B}(\kappa)$ be the above Boolean algebra. Then we have the following Boolean values in $V^{\mathcal{B}(\kappa)}$:

(i) $[\![\text{'}|(P(\kappa))^{\vee}| = |(\kappa^{+})^{\vee}|\text{'}]\!] = 1$;

(ii) $[\![\text{'}P(\check{\kappa}) = (P(\kappa))^{\vee}\text{'}]\!] = 1$;

(iii) $[\![\text{'}|P(\check{\kappa})| = |(\kappa^{+})^{\vee}|\text{'}]\!] = 1$;

(iv) $[\![\text{'}|\check{\kappa}| = \check{\kappa}\text{'}]\!] = 1$;

(v) $[\![\text{'}2^{\check{\kappa}} = (\check{\kappa})^{+}\text{'}]\!] = 1$.

<u>Proof</u>

Set $\lambda = \kappa^{+}$.

(i) Define $F : \lambda \times P(\kappa) \to B(\kappa)$ by

$$F(\alpha,\sigma) = \bigvee\{f \in Q(\kappa) : \alpha \in \operatorname{dom} f,\ f(\alpha) = \alpha\}.$$

It is easily checked that F satisfies the hypotheses (i) - (iv) of 7.26 (with $b = 1$) : for example, for $\sigma \subset \kappa$,

$$\bigvee\{F(\alpha,\sigma) : \alpha < \kappa\} =$$
$$\bigvee\{f \in Q(\kappa) : f(\alpha) = \sigma \text{ for some } \alpha \in \operatorname{dom} f\} = 1$$

because $\{f \in Q(\kappa) : \sigma \in \operatorname{ran} f\}$ is dense in $(Q(\kappa),\leq)$. So, by 7.26, $[\![\text{'}|(P(\kappa))^{\vee}| = |\check{\lambda}|\text{'}]\!] = 1$.

(ii) Let $b = [\![\text{'}P(\check{\kappa}) \neq (P(\kappa))^{\vee}\text{'}]\!]$. By the fullness lemma, there exists $v \in V^{\mathcal{B}(\kappa)}$ with

$$[\![v \subset \check{\kappa}]\!] \wedge [\![v \hat{\notin} (P(\kappa))^{\vee}]\!] = b.$$

Suppose, if possible, that $b \neq 0$. Then we *claim* that there is a sequence $\langle f_\alpha : \alpha \leq \kappa \rangle \subset Q(\kappa)$ such that:

(a) $f_\beta \leq f_\alpha \leq b \quad (\alpha \leq \beta < \kappa)$;

(b) for $\alpha < \kappa$, either $f_\alpha \leq [\![\check{\alpha} \hat{\in} v]\!] \wedge b$ or $f_\alpha \leq [\![\check{\alpha} \hat{\notin} v]\!] \wedge b$.

For suppose that $\alpha \leq \kappa$ and that $\langle f_\beta : \beta < \alpha \rangle$ has been constructed. Set $f = \bigcup\{f_\beta : \beta < \alpha\}$. Then f is a well-defined, injective map from $\bigcup\{\text{dom } f_\beta : \beta < \alpha\}$ into $P(\kappa)$. Since $\lambda = \kappa^+$ is a regular cardinal (page 87), $\text{dom } f \in \kappa^+$. Hence $f \in Q(\kappa)$. Since $Q(\kappa)$ is dense in $B(\kappa)\setminus\{0\}$, there exists $f_\alpha \in Q(\kappa)$ with $f_\alpha \leq f$ and with either

$$f_\alpha \leq [\![\check{\alpha} \hat{\in} v]\!] \wedge b \quad \text{or} \quad f_\alpha \leq [\![\check{\alpha} \hat{\notin} v]\!] \wedge b.$$

The construction continues, and the claim is established.

Set $f = f_\kappa$, and set

$$\sigma = \{\alpha < \kappa : f_\alpha \leq [\![\check{\alpha} \hat{\in} v]\!]\}.$$

Then $f \leq b$ and

$$f \leq [\![\check{\alpha} \hat{\in} v]\!] \quad (\alpha \in \sigma), \qquad f \leq [\![\check{\alpha} \hat{\notin} v]\!] \quad (\alpha \in \kappa\setminus\sigma).$$

Since $[\![v \subset \check{\kappa}]\!] \geq b$,

$$[\![v \subset \check{\sigma}]\!] \wedge b = \bigwedge\{[\![\check{\alpha} \hat{\in} v]\!] \wedge b : \alpha \in \kappa\setminus\sigma\}.$$

Also $[\![\check{\sigma} \subset v]\!] = \bigwedge\{[\![\check{\alpha} \hat{\in} v]\!] : \alpha \in \sigma\}$, and so

$$f \leq [\![\check{\sigma} \subset v]\!] \wedge [\![v \subset \check{\sigma}]\!] = [\![v \hat{=} \hat{\sigma}]\!]$$

$$\leq [\!['v \in P(\kappa)^{\vee}']\!] \leq b'.$$

Thus $f \leq b \wedge b' = 0$, a contradiction. Thus $b = 0$, and (ii) follows.

(iii) This is immediate from (i) and (ii).

(iv) If the result does not hold, then there exists $\alpha < \kappa$ and a term $t \in V^{\mathcal{B}(\kappa)}$ such that

$$[\![\text{'}t \text{ is a map from } \check{\alpha} \text{ onto } \check{\kappa}\text{'}]\!] \geq b \neq 0,$$

and this implies that $[\![\text{'}t \subset \check{\alpha} \times \check{\kappa}\text{'}]\!] \geq b$. Now, from (ii),

$$[\![\text{'}P(\check{\alpha} \times \check{\kappa}) = (P(\alpha \times \kappa))^{\vee}\text{'}]\!] = 1,$$

and so there exists $s \subset \alpha \times \kappa$ such that $[\![t \hat{=} \check{s}]\!] \wedge b \neq 0$. Thus

$$[\![\text{'}\check{s} \text{ is a map from } \check{\alpha} \text{ onto } \check{\kappa}\text{'}]\!] > 0.$$

It is now easy to check (cf. 7.24(iii)) that $s : \alpha \to \kappa$ is a surjection, a contradiction of the fact that $\alpha < \kappa$.

(v) By (iii), $[\![\text{'}2^{\check{\kappa}} = |(\kappa^+)^{\vee}|\text{'}]\!] = 1$. By (i), $[\![\text{'}\check{\kappa} \text{ is a cardinal'}]\!] = 1$, and so $[\![\text{'}|(\kappa^+)^{\vee}| \leq (\check{\kappa})^+\text{'}]\!]$. The result follows. ∎

7.37 THEOREM

There is a complete Boolean algebra $\mathcal{B}$ such that $[\![\text{CH}]\!] = 1$ in $V^{\mathcal{B}}$.

<u>Proof</u>

Take $\mathcal{B} = \mathcal{B}(\aleph_0)$ in the above notation. By 7.36(v),

$$[\![\text{'}2^{\check{\aleph}_0} = \check{\aleph}_1\text{'}]\!] = 1 \text{ in } V^{\mathcal{B}}. \quad ∎$$

7.38 THEOREM

Assume that there is a model $\mathfrak{M}$ of ZFC. Then there is a model, extending $\mathfrak{M}$, of ZFC + CH. ∎

We have given a full proof, modulo Gödel's completeness theorem, that CH is independent of ZFC.

7.39 THEOREM (CH)

There is a complete Boolean algebra $\mathcal{B}$ such that

$$[\![\text{'} 2^{\aleph_0} = \aleph_1 \text{ and } 2^{\aleph_1} = \aleph_2 \text{'}]\!] \text{ in } V^{\mathcal{B}}.$$

<u>Proof</u>

Take $\mathcal{B} = \mathcal{B}(\aleph_1)$ in the above notation. By 7.36(iv), $[\![\text{'}\check{\aleph}_1 = \aleph_1\text{'}]\!] = 1$, and, by 7.36(v), $[\![\text{'}2^{\aleph_1} = \aleph_2\text{'}]\!] = 1$.

By 7.36(ii), $[\![\text{'}P(\check{\aleph}_1) = (P(\aleph_1))^{\vee}\text{'}]\!] = 1$, and so $[\![\text{'}|P(\check{\mathbb{N}})| = |P(\mathbb{N})|^{\vee}\text{'}]\!] = 1$. But, by CH, $|P(\mathbb{N})| = \aleph_1$, and so $[\![\text{'}|P(\mathbb{N}))^{\vee}| = |\check{\aleph}_1|\text{'}]\!] = 1$. Hence $[\![\text{'}|P(\check{\mathbb{N}})| = |\check{\aleph}_1|\text{'}]\!] = 1$. But $[\![\text{'}|\check{\aleph}_1| \leqslant \aleph_1\text{'}]\!] = 1$, and so $[\![\text{'}|P(\check{\mathbb{N}})| \leqslant \aleph_1\text{'}]\!] = 1$, i.e., $[\![\text{'}2^{\aleph_0} = \aleph_1\text{'}]\!] = 1$. ∎

The remaining topic in this chapter is the delivery of the promised proof that $[\![\text{AC}]\!] = 1$. We begin by introducing a notation for an already familiar construction. The notation will be used frequently in Chapter 8.

7.40 DEFINITION

Let $\mathcal{B}$ be a complete Boolean algebra, and let S be a subset of $V^{\mathcal{B}}$. Set

$$\text{dom } S^{\#} = S, \quad S^{\#}(t) = 1 \quad (t \in S).$$

Thus $S^{\#}$ is a term in $V^{\mathcal{B}}$. For example, if $u \in V$, then $\check{u} = (\text{dom } \check{u})^{\#}$.

Note that, for each $t \in V^{\mathcal{B}}$,

$$[\![\text{'}t \subset (\text{dom } t)^{\#}\text{'}]\!] = 1. \tag{13}$$

We must now contend with the somewhat messy problem of explicitly constructing terms for functions.

For $s, t \in V^{\mathcal{B}}$, set $w = \{\{s\}^{\#}, \{s,t\}^{\#}\}^{\#}$. It is routine to check that $[\![\text{'}w = (s,t)\text{'}]\!] = 1$. (Recall that, if w = (s,t), then, formally, w = {{s},{s,t}}.)

Now suppose that S is a subset of $V^{\mathcal{B}}$ and that $f : S \to V^{\mathcal{B}}$ is a function. Set

$$\mathit{f} = \{\{\{s\}^{\#},\{s,f(s)\}^{\#}\}^{\#} : s \in S\}^{\#}.$$

Then $[\![$ 'f is a binary relation' $]\!] = 1$. Next, suppose that f has the additional property that

$$[\![s \mathrel{\hat{=}} t]\!] \leq [\![f(s) \mathrel{\hat{=}} f(t)]\!] \quad (s,t \in S). \tag{14}$$

Then $[\![$ 'f is a function with domain $S^{\#}$' $]\!] = 1$, and, for each $s \in S$,

$$[\![\text{'the value of } \mathit{f} \text{ at } s \text{ is } f(s)\text{'}]\!] = 1.$$

Thus

$$[\![\text{'}\mathit{f} \text{ is a function from } S^{\#} \text{ onto } f(S)^{\#}\text{'}]\!] = 1.$$

Finally, if f satisfies the condition that

$$[\![s \mathrel{\hat{=}} t]\!] = [\![f(s) \mathrel{\hat{=}} f(t)]\!] \quad (s,t \in S), \tag{15}$$

then

$$[\![\text{'}\mathit{f} \text{ is a bijection from } S^{\#} \text{ onto } f(S)^{\#}\text{'}]\!] = 1.$$

7.41 THEOREM

Let $\mathcal{B}$ be a complete Boolean algebra. Then $[\![AC]\!] = 1$ in $V^{\mathcal{B}}$.

<u>Proof</u>

By 5.4, AC is equivalent (in ZF) to the sentence ϕ_o:

'for each set x, there is a function y such that dom y is an ordinal and x ⊂ ran y'.

By 7.13, $[\![\phi]\!] = 1$ in $V^{\mathcal{B}}$ for each axiom ϕ of ZF, and so it suffices to show that $[\![\phi_o]\!] = 1$.

Let $u \in V^{\mathcal{B}}$. By passing to an equivalent term, we may suppose that $\mathrm{dom}\, u \neq \emptyset$. By the Axiom of Choice, there is an ordinal α and a function h from α onto dom u. Set $S = \{\check{\beta} : \beta < \alpha\}$, define $f : S \to V^{\mathcal{B}}$ by $f(\check{\beta}) = h(\beta)$ $(\beta < \alpha)$, and let f be the corresponding term, as above.

Take $s,t \in S$. By 7.18(ii), $s \neq t$ if and only if $[\![s \hat{=} t]\!] = 0$, and so $[\![s \hat{=} t]\!] \leqslant [\![f(s) \hat{=} f(t)]\!]$. Since $S^{\#} = \check{\alpha}$, it follows that

$$[\![\text{'}f \text{ is a function from } \check{\alpha} \text{ onto } (\mathrm{dom}\, u)^{\#}\text{'}]\!] = 1.$$

By 7.20(ii), $[\![\text{'}\check{\alpha}$ is an ordinal' $]\!] = 1$, and so, using (13) it follows that f has the required properties. Thus $[\![\phi_o]\!] = 1$ in $V^{\mathcal{B}}$. ∎

7.42 NOTES

We have remarked that the relative consistency of ZFC + ¬CH was an open question for 25 years after Gödel's proof in 1938 of the relative consistency of ZFC + CH. The proof of Cohen was given in [11] and [12] and was expounded in [13]. The approach to Cohen's method through $V^{\mathcal{B}}$ was discovered by Solovay in 1965, and is given in [60]. See [59, §6] and the forward to the second edition of [6] for some interesting historical remarks by Scott. An early use of Boolean-valued models to give a proof of Gödel's completeness theorem is due to Rasiowa and Sikorski ([55]).

Accounts of forcing are given in the standard texts [6], [44], [49], and [62], and in the article of Burgess [9].

We proved that it is relatively consistent with ZFC that $2^{\aleph_o} = \aleph_2$. In fact, it is relatively consistent that the values of 2^{κ} for κ a regular cardinal can be any collection of cardinals subject to: (i) $\mathrm{cf}\,(2^{\kappa}) > \kappa$; (ii) if $\kappa \leqslant \lambda$, then $2^{\kappa} \leqslant 2^{\lambda}$. See [44, Theorem 46] and [49, VIII.4]. The behaviour of 2^{κ} at singular cardinals κ is still not completely resolved; some recent results are given in [33].

8 ITERATED FORCING

We shall now complete our proof of the relative consistency with ZFC of the assertion that each norm on each algebra $C(X,\mathbb{C})$ is equivalent to the uniform norm. To do this, we shall construct a complete Boolean algebra $\mathbb{A}$ with

$$[\![MA + NDH]\!] = 1 \text{ in } V^{\mathbb{A}}.$$

We recall that MA is Martin's Axiom (Definition 5.14) and that NDH is the sentence "For each compact space X, each homomorphism from $C(X,\mathbb{C})$ into a Banach algebra is continuous" (Definition 4.18). The existence of discontinuous homomorphism from $C(X,\mathbb{C})$ is equivalent to the existence of norms on $C(X,\mathbb{C})$ which are not equivalent to the uniform norm. We have explained how the construction of such a Boolean algebra gives our consistency result.

In this chapter, we write $[\![\ldots]\!]^{\mathbb{B}}$ for a Boolean value in $V^{\mathbb{B}}$. Let us discuss how we might construct a Boolean algebra $\mathbb{A}$ with $[\![MA]\!]^{\mathbb{A}} = 1$ in $V^{\mathbb{A}}$.

Suppose, for example, that $\mathbb{B}$ is a complete Boolean algebra and that t is a term in $V^{\mathbb{B}}$ which is a counter-example to MA in $V^{\mathbb{B}}$, in the sense that

$$[\![\text{'}t \text{ is a counter-example to MA'}]\!]^{\mathbb{B}} = 1.$$

Then we shall show that there exists a complete Boolean algebra $\mathbb{C}$ containing $\mathbb{B}$ as a complete subalgebra such that

$$[\![\text{'}t \text{ is a counter-example to MA'}]\!]^{\mathbb{C}} = 0.$$

In this way, we can eliminate the counter-example t in $V^{\mathbf{C}}$. The hope is that, by iterating this process, we can exterminate all possible counter-examples to MA. Of course, new counter-examples could well arise at each stage of the iteration, and these must also be eliminated. This is how we proceed. The final algebra $\mathbf{A}$ will essentially be constructed as a union of subalgebras, and will have all the required properties.

We shall arrange that $[\![\mathrm{NDH}]\!] = 1$ in $V^{\mathbf{A}}$ by "folding in" certain extra properties as the iteration continues.

In preparation for this construction, we must develop a theory of "forcing iterations" to an extent sufficient to prove two iteration theorems, 8.12 and 8.14. This development is presented in pp.184-205. However, the iteration theorems are stated in such a way that the construction of the Boolean algebra $\mathbf{A}$, to be given in the second half of this chapter, can be understood from the statements of the two theorems, without additional reference to the preliminary results.

We shall comment briefly on the history of the results of this chapter in the notes 8.27, but we should say here that the initial development of iterated forcing is due to Solovay and Tennenbaum ([67]). All our basic results about iterations are contained in [67], but our details differ: we are much more explicit, and, of course, we use a less sophisticated language.

Throughout this chapter we shall frequently use the equivalence relation $\sim$, defined on page 143: for $u,v \in V^{\mathbf{B}}$,

$$u \sim v \text{ if and only if } [\![u \mathrel{\hat{=}} v]\!]^{\mathbf{B}} = 1.$$

8.1 DEFINITION

Let $\mathbf{B}$ and $\mathbf{C}$ be complete Boolean algebras such that $\mathbf{B}$ is a subalgebra of $\mathbf{C}$. Then $\mathbf{B}$ is a <u>complete subalgebra</u> of $\mathbf{C}$ if, for each $S \subset B$, the suprema of S

calculated in $\mathcal{B}$ and in $\mathcal{C}$ are equal.

To show that $\mathcal{B}$ is a complete subalgebra of $\mathcal{C}$, it suffices to check that, for each subset S of B with $\bigvee S = 1$ in $\mathcal{B}$, $\bigvee S = 1$ in $\mathcal{C}$.

For example, the trivial Boolean algebra (2) is a complete subalgebra of each complete Boolean algebra.

Let $\mathcal{B}$ be a complete subalgebra of $\mathcal{C}$. Then $V^{\mathcal{B}} \subset V^{\mathcal{C}}$, and it is routine to check that, for $s, t \in V^{\mathcal{B}}$, $[\![s \mathrel{\hat{\in}} t]\!]^{\mathcal{B}} = [\![s \mathrel{\hat{\in}} t]\!]^{\mathcal{C}}$ and $[\![s \mathrel{\hat{=}} t]\!]^{\mathcal{B}} = [\![s \mathrel{\hat{=}} t]\!]^{\mathcal{C}}$. The rank of a term in $V^{\mathcal{B}}$ is the same as its rank in $V^{\mathcal{C}}$.

Many of the basic properties of terms are preserved in the passage from $V^{\mathcal{B}}$ to $V^{\mathcal{C}}$. This is exactly analogous to the case of the passage from V to $V^{\mathcal{B}}$ discussed in Chapter 7. We now give the general theorem which asserts that properties specified by certain formulae are preserved.

8.2 DEFINITION

The set of <u>Δ_o-formulae in $\mathcal{L}$</u> is the smallest set Φ of formulae of $\mathcal{L}$ such that: (i) Φ contains the atomic formulae; (ii) Φ contains $\phi \vee \psi$ and $\neg\phi$ whenever Φ contains ϕ and ψ; (iii) Φ contains $\exists x_i(x_i \mathrel{\hat{\in}} x_j \wedge \phi)$ whenever Φ contains ϕ and x_i and x_j are variables.

Many simple properties of sets can (possibly with some effort) be specified by a Δ_o-formula. For example, $x \subset y$, which is really $\forall z(z \mathrel{\hat{\in}} x \Rightarrow z \mathrel{\hat{\in}} y)$, is logically equivalent to $\neg(\exists z(z \mathrel{\hat{\in}} x \wedge \neg(z \mathrel{\hat{\in}} y)))$, which is a Δ_o-formula.

Properties which can be formalized by Δ_o-formulae include 'x is a function' and 'x is a Boolean algebra' (but *not* 'x is a complete Boolean algebra'). A list of such properties is verified in [49, IV.3.9], and we shall later give a list relevant to our discussion.

Let $\phi(x_1,\ldots,x_n)$ be a Δ_o-formula.

8.3 PROPOSITION

Let $\mathcal{B}$ be a complete subalgebra of $\mathcal{C}$, and let

$t_1,\ldots,t_n \in V^{\mathbf{B}}$. Then

$$[\![\phi[t_1,\ldots,t_n]]\!]^{\mathbf{B}} = [\![\phi[t_1,\ldots,t_n]]\!]^{\mathbf{C}}.$$

Proof

The proof is by induction on the length of subformulae of ϕ.

Suppose, for example, that the result holds for a subformula $\psi(x,y)$ of ϕ, that $\chi(y)$ is $\exists x(x \mathbin{\hat\in} y \wedge \psi(x,y))$, and that χ is a subformula of ϕ. For $t \in V^{\mathbf{B}}$,

$$\begin{aligned}[\![\chi[t]]\!]^{\mathbf{B}} &= \bigvee\{[\![s \mathbin{\hat\in} t]\!]^{\mathbf{B}} \wedge [\![\psi[s,t]]\!]^{\mathbf{B}} : s \in \mathrm{dom}\, t\}\\ &= \bigvee\{[\![s \mathbin{\hat\in} t]\!]^{\mathbf{C}} \wedge [\![\psi[s,t]]\!]^{\mathbf{C}} : s \in \mathrm{dom}\, t\}\\ &= [\![\chi[t]]\!]^{\mathbf{C}},\end{aligned}$$

and so the result holds for χ. The remainder of the proof is routine. ∎

Let $u \in V^{(2)}$, and set $v = \pi(u)$, where π was defined on page 138. It is easily proved by induction on rank that $u \sim \check{v}$ (and so V is "isomorphic to $V^{(2)}/\sim$"). Thus, by 8.3, properties expressed by Δ_o-formulae are also preserved in the passage from V to $V^{\mathbf{B}}$ (cf. page 167) because they are preserved in the passage from $V^{(2)}$ to $V^{\mathbf{B}}$.

We now begin our discussion of iteration. Let $\mathbf{B}$ be a complete Boolean algebra, and define a term G in $V^{\mathbf{B}}$ by

$$\mathrm{dom}\, G = \{\check{b} : b \in B\}, \quad G(\check{b}) = b \quad (b \in B).$$

Then $[\![\check{b} \mathbin{\hat\in} G]\!]^{\mathbf{B}} = b \quad (b \in B)$ and $[\![G \subset \check{B}]\!]^{\mathbf{B}} = 1$. We *claim* that

$$[\![\text{'}G \text{ is an ultrafilter in } \check{B}\text{'}]\!]^{B} = 1.$$

To see this, note that $[\![\text{'}\check{B} \text{ is a Boolean algebra'}]\!]^{B} = 1$ by the above remarks on Δ_o-formulae. We then check that all of the defining properties of the statement 'G is an ultrafilter in $\check{B}$' have value 1 in V^{B}.

8.4 DEFINITION

The term G is the term for the generic object in V^{B}.

Recall that a subset D of $B\setminus\{0\}$ is dense in B if, for each $a \in B\setminus\{0\}$, there exists $d \in D$ with $d \leqslant a$, and that a subset F of $B\setminus\{0\}$ is $\mathcal{D}$-generic, where $\mathcal{D} \subset P(B)$, if $F \cap D \neq \emptyset$ $(D \in \mathcal{D})$.

The following result gives the key property of G (and it essentially characterizes G).

8.5 PROPOSITION

Let G be the term for the generic object in V^{B}, and let $\mathcal{D} = \{D \subset B\setminus\{0\} : D \text{ is dense in } B\}$. Then

$$[\![\text{'}G \text{ is } \check{\mathcal{D}}\text{-generic'}]\!]^{B} = 1.$$

Proof

Let $D \in \mathcal{D}$. Then $[\![\text{'}\check{D} \text{ is dense in } \check{B}\text{'}]\!]^{B} = 1$, and

$$[\![\text{'}\check{D} \cap G \neq \emptyset\text{'}]\!]^{B} = \bigvee\{[\![\check{d} \,\hat{\in}\, G]\!]^{B} : d \in D\} = \bigvee D = 1$$

because D is dense. This implies the result. ∎

Let F be a prefilter in a Boolean algebra B (so that F is a non-empty subset of $B\setminus\{0\}$ such that, if $a,b \in F$, then there exists $c \in F$ with $c \leqslant a$ and $c \leqslant b$). For $a,b \in B$, set $a \sim_F b$ if $a \wedge c = b \wedge c$ for some $c \in F$. Then $\sim_F$ is an equivalence relation on B. The equivalence class containing a is denoted by $[a]_F$, and $B/\sim_F$ with the operations $[a]_F \wedge [b]_F = [a \wedge b]_F$,

$[a]_F \vee [b]_F = [a \vee b]_F$ is a Boolean algebra; it is denoted by $\mathbb{B}/\sim_F$.

Let $\mathbb{B}$ be a complete Boolean algebra. The theory of iterations that we are to give depends on an analysis of terms u in $V^{\mathbb{B}}$ such that

$$[\![\text{'}u \text{ is a complete Boolean algebra'}]\!]^{\mathbb{B}} = 1.$$

Let $\mathbb{C}$ be a complete Boolean algebra containing $\mathbb{B}$ as a complete subalgebra, and let G be the term for the generic object in $V^{\mathbb{B}}$. Since

$$[\![\text{'}G \text{ is an ultrafilter in } \check{\mathbb{B}}\text{'}]\!]^{\mathbb{B}}$$

$$= [\![\text{'}\check{\mathbb{B}} \text{ is a subalgebra of } \check{\mathbb{C}}\text{'}]\!]^{\mathbb{B}} = 1,$$

we have $[\![\text{'}G \text{ is a prefilter in } \check{\mathbb{C}}\text{'}]\!]^{\mathbb{B}} = 1$, and so

$$[\![\text{'}\check{\mathbb{C}}/\sim_G \text{ is a Boolean algebra'}]\!]^{\mathbb{B}} = 1.$$

For $c \in C$, define $[c]_G \in V^{\mathbb{B}}$ by

$$\left.\begin{aligned} &\mathrm{dom}[c]_G = \{\check{s} : s \in C\}, \\ &[c]_G(\check{s}) = \bigvee\{b \in B : b \wedge c = b \wedge s\} \quad (s \in C). \end{aligned}\right\} \quad (1)$$

Let us stress that $[c]_G$ is a term (and not a subset of C). We have

$$[\![\text{'}[c]_G \text{ is the equivalence class containing } \check{c} \text{ in } \check{C}/\sim_G\text{'}]\!]^{\mathbb{B}} = 1.$$

We list the basic properties of the map $c \mapsto [c]_G$.

8.6 LEMMA

Let $b \in B$ and let $c,d \in C$.

(i) $[\![\text{'}[c \vee d]_G = [c]_G \vee [d]_G \text{ in } \check{C}/\sim_G\text{'}]\!]^{B} = 1$,
etc.

(ii) $b \leq [\![[c]_G \hat{=} [d]_G]\!]^{B}$ if and only if
$b \wedge c = b \wedge d$.

(iii) $[\![[c]_G \hat{=} [d]_G]\!]^{B} = 1$ if and only if $c = d$.

(iv) $b \leq [\![\text{'}[c]_G \leq [d]_G \text{ in } \check{C}/\sim_G\text{'}]\!]^{B}$ if and only if $b \wedge c \leq b \wedge d$.

(v) If $t \in V^{B}$ with $[\![\text{'}t \in \check{C}/\sim_G\text{'}]\!]^{B} = 1$, then there exists $c \in C$ with $t \sim [c]_G$.

(vi) Suppose that $S \subset C$, that $c = \bigvee S$, and that $T = \{[d]_G : d \in S\}$. Then

$$[\![\text{'}[c]_G = \bigvee T^{\#} \text{ in } \check{C}/\sim_G\text{'}]\!]^{B} = 1.$$

<u>Proof</u>

For (v), set $c = \bigvee\{s \wedge [\![[s]_G \hat{=} t]\!]^{B} : s \in C\}$.
The remaining results are immediate from the definition. ∎

8.7 LEMMA

With the above notation,

$$[\![\text{'}\check{C}/\sim_G \text{ is a complete Boolean algebra'}]\!]^{B} = 1.$$

<u>Proof</u>

Take $t \in V^{B}$ such that $[\![\text{'}t \subset \check{C}/\sim_G\text{'}]\!]^{B} = 1$, and set $c = \bigvee\{[\![[d]_G \hat{\in} t]\!]^{B} \wedge d : d \in C\}$. Then it is routine to check that $[\![\text{'}[c]_G = \bigvee t \text{ in } \check{C}/\sim_G\text{'}]\!]^{B} = 1$. ∎

Before proceeding, we recall from page 181 that, if $f : S \to V^{B}$ is a function such that

$$[\![s \hat{=} t]\!]^{B} = [\![f(s) \hat{=} f(t)]\!]^{B} \quad (s,t \in S), \qquad (2)$$

then $[\![\text{'}f \text{ is a bijection from } S^{\#} \text{ onto } f(S)^{\#}\text{'}]\!]^{B} = 1$, where f is the term corresponding to f.

8.8 THEOREM

(i) Let $\mathbb{B}$ be a complete Boolean algebra, and let $u \in V^{\mathbb{B}}$ be a term such that

$$[\![\text{'}u \text{ is a complete Boolean algebra'}]\!]^{\mathbb{B}} = 1.$$

Then there is a complete Boolean algebra $\mathbb{C}$ containing $\mathbb{B}$ as a complete subalgebra such that

$$[\![\text{'}u \text{ is isomorphic to } \check{\mathbb{C}}/\sim_G \text{'}]\!]^{\mathbb{B}} = 1.$$

(ii) Suppose, further, that $\mathbb{B}$ is ccc and has cardinality $2^{\aleph_0}$ and that

$$[\![\text{'}u \text{ is ccc and has cardinality } \leqslant 2^{\aleph_0}\text{'}]\!]^{\mathbb{B}} = 1.$$

Then we can suppose that $\mathbb{C}$ is ccc and has cardinality $2^{\aleph_0}$.

<u>Proof</u>

We write $[\![\ldots]\!]$ for $[\![\ldots]\!]^{\mathbb{B}}$ in this proof.

By the fullness lemma 7.12(ii), there exists $v \in V^{\mathbb{B}}$ with

$$[\![\text{'}v \text{ is the underlying set of } u\text{'}]\!] = 1.$$

Take $\alpha = \rho(v) + 1$. The restriction of $\sim$ to

$$\{s \in V_\alpha^{\mathbb{B}} : [\![s \mathbin{\hat{\in}} v]\!] = 1\}$$

is an equivalence relation on a set: form a subset C^* of $V_\alpha^{\mathbb{B}}$ by choosing exactly one element from each equivalence class. Then:

(a) if $s, t \in C^*$ with $s \neq t$, then $[\![s \mathbin{\hat{=}} t]\!] \neq 1$.

Let $t \in V^{\mathbb{B}}$ with $[\![t \mathbin{\hat{\in}} v]\!] = 1$, and define $r \in V^{\mathbb{B}}$ by

$$\text{dom } r = V^{\mathbf{B}}_{\rho(v)}, \quad r(w) = [\![w \,\hat{\in}\, t]\!] \quad (w \in \text{dom } r).$$

Clearly $[\![r \subset t]\!] = 1$. Since $[\![t \,\hat{\in}\, v]\!] = 1$, we have

$$[\![w \,\hat{\in}\, t]\!] \leqslant \bigvee\{[\![w \,\hat{\in}\, s]\!] : s \in \text{dom } v\}$$
$$\leqslant \bigvee\{[\![w \,\hat{\in}\, s]\!] : s \in V^{\mathbf{B}}_{\rho(v)}\} \leqslant \bigvee\{[\![w \,\hat{=}\, s]\!] : s \in V^{\mathbf{B}}_{\rho(v)}\}$$

for $w \in V^{\mathbf{B}}$, and so $[\![t \subset r]\!] = 1$. Hence $r \sim t$. Further, $r \in V^{\mathbf{B}}_{\alpha}$, and so:

(b) if $t \in V^{\mathbf{B}}$ with $[\![t \,\hat{\in}\, v]\!] = 1$, then there exists $s \in C^*$ with $s \sim t$.

Define operations $\wedge$ and $\vee$ on C^* by the rules

$$s_1 \wedge s_2 = s_3 \quad \text{if} \quad [\![\psi_\wedge(s_1, s_2, s_3, u)]\!] = 1,$$
$$s_1 \vee s_2 = s_3 \quad \text{if} \quad [\![\psi_\vee(s_1, s_2, s_3, u)]\!] = 1,$$

where $\psi_\wedge$ and $\psi_\vee$ were defined on page 155. Thus $s_1 \wedge s_2 = s_3$ if $[\![\text{'}s_1 \wedge s_2 = s_3 \text{ in } u\text{'}]\!] = 1$. The operations are well defined, and C^*, together with these operations, is a Boolean algebra, to be denoted by $\mathfrak{C}^*$.

Let $b \in B$. By the mixing lemma, there exists $b^* \in V^{\mathbf{B}}$ with

$$[\![\text{'}b^* = 1 \text{ in } u\text{'}]\!] \geqslant b \quad \text{and}$$
$$[\![\text{'}b^* = 0 \text{ in } u\text{'}]\!] \geqslant b'.$$

Since $[\![b^* \,\hat{\in}\, v]\!] = 1$, we may, by (b), suppose that $b^* \in C^*$, and then b^* is uniquely specified by these properties.

Set $B^* = \{b^* : b \in B\}$. Then $\mathbf{B}^*$ is a Boolean subalgebra of $\mathfrak{C}^*$; the map $b \mapsto b^*$, $\mathbf{B} \to \mathbf{B}^*$, is a Boolean isomorphism, and so $\mathbf{B}^*$ is complete.

We first *claim* that $\mathfrak{C}^*$ is complete.

Let S be a subset of C^*, so that $[\![s \,\hat{\in}\, v]\!] = 1$ $(s \in S)$ and $[\![S^\# \subset v]\!] = 1$. By the fullness lemma 7.12(ii), there exists $c \in V^{\mathbf{B}}$ such that $[\![\text{'}c = \bigvee S^\# \text{ in } u\text{'}]\!] = 1$, and, by (b), we may suppose that $c \in C^*$. For $s \in S$, $[\![s \mathrel{\hat{\in}} S^\#]\!] \geqslant S^\#(s) = 1$, and so $s \leqslant c$ in C^*. Now suppose that $d \in C^*$ and that $s \leqslant d$ $(s \in S)$. Then

$$[\![\text{'}c \leqslant_u d\text{'}]\!] = [\![\text{'}s \leqslant_u d \quad (s \in S^\#)\text{'}]\!]$$

$$= \bigwedge\{[\![\text{'}s \leqslant_u d\text{'}]\!] : s \in S\} = 1,$$

and so $c \leqslant d$. This establishes that $c = \bigvee S$ in $\mathfrak{C}^*$, and so $\mathfrak{C}^*$ is complete.

Secondly, we *claim* that $\mathfrak{B}^*$ is a complete subalgebra of $\mathfrak{C}^*$.

Take $T \subset B$ with $\bigvee T = 1$ in $\mathfrak{B}$ and $c = \bigvee\{b^* : b \in T\}$ in $\mathfrak{C}^*$. Clearly $[\![\text{'}b^* \leqslant_u c\text{'}]\!] = 1$ $(b \in T)$, and so for $b \in T$,

$$[\![\text{'}c = 1 \text{ in } u\text{'}]\!] \geqslant [\![\text{'}b^* \leqslant_u c\text{'}]\!] \wedge [\![\text{'}b^* = 1 \text{ in } u\text{'}]\!]$$

$$= [\![\text{'}b^* = 1 \text{ in } u\text{'}]\!] \geqslant b.$$

Thus $[\![\text{'}c = 1 \text{ in } u\text{'}]\!] = 1$, and so $c = 1$. This is sufficient to establish the second claim.

We now choose a complete Boolean algebra $\mathfrak{C}$ containing $\mathfrak{B}$ as a complete subalgebra such that there is a Boolean isomorphism $\pi: \mathfrak{C} \to \mathfrak{C}^*$ with $\pi(b) = b^*$ $(b \in B)$. We finally *claim* that

$$[\![\text{'}u \text{ is isomorphic to } \check{\mathfrak{C}}/\sim_G\text{'}]\!] = 1.$$

Take $c_1, c_2 \in C$ and $b \in B$. Then $b \wedge c_1 = b \wedge c_2$ if and only if $b^* \wedge \pi(c_1) = b^* \wedge \pi(c_2)$ in $\mathfrak{C}^*$, and so, by 8.6(iv), $b \leqslant [\![[c_1]_G \mathrel{\hat{=}} [c_2]_G]\!]$ if and only if

$$[\![\text{'}b^* \wedge \pi(c_1) = b^* \wedge \pi(c_2) \text{ in } u\text{'}]\!] = 1. \qquad (3)$$

Since $[\![\text{'}b^* = 1 \text{ in } u\text{'}]\!] \geqslant b$, (3) holds if and only if

$[\![\pi(c_1) \mathrel{\hat{=}} \pi(c_2)]\!] \geq b$. Thus

$$[\![[c_1]_G \mathrel{\hat{=}} [c_2]_G]\!] = [\![\pi(c_1) \mathrel{\hat{=}} \pi(c_2)]\!]. \qquad (4)$$

Set $S = \{[c]_G : c \in C\}$ and $f : [c]_G \mapsto \pi(c)$, $S \to D$; f is well defined, for if $[c_1]_G = [c_2]_G$, then $c_1 = c_2$ by 8.6(iii). By 8.6(i), f is a homomorphism. By (4), f satisfies (2), and so

$$[\![\text{'}\mathit{f} \text{ is an isomorphism from } \check{\mathfrak{C}}/\sim_G \text{ onto } \mathit{u}\text{'}]\!] = 1,$$

where f is the term corresponding to f. Thus the final claim holds, and (i) is proved.

(ii) We write $\mathfrak{c}$ for $2^{\aleph_0}$.

We prove that $\mathfrak{C}^*$ is ccc and has cardinality $\mathfrak{c}$: this is sufficient.

Let S be an antichain in $\mathfrak{C}^*$, so that $S \subset C^*\setminus\{0\}$ and $s \wedge t = 0$ for $s,t \in S$ with $s \neq t$. Then

$$[\![\text{'}S^\#\setminus\{0\} \text{ is an antichain in } \mathit{u}\text{'}]\!] = 1.$$

Since $[\![\text{'}\mathit{u} \text{ is ccc'}]\!] = 1$, it follows that

$$[\![\text{'}S^\#\setminus\{0\} \text{ is countable'}]\!] = 1.$$

By the fullness lemma, there exists $t \in V^{\mathbf{B}}$ with

$$[\![\text{'}t \text{ is a surjection from } \check{\mathbb{N}} \text{ onto } S^\#\setminus\{0\}\text{'}]\!] = 1.$$

Take $n \in \mathbb{N}$. Then there exists $t_n \in V^{\mathbf{B}}$ with

$$[\![\text{'the value of } t \text{ at } \check{n} \text{ is } t_n\text{'}]\!] = 1.$$

For $s,t \in S$ with $s \neq t$, $[\![\text{'}s \wedge t = 0 \text{ in } \mathit{u}\text{'}]\!] = 1$, and so

$$[\![t_n \mathrel{\hat{=}} s]\!] \wedge [\![t_n \mathrel{\hat{=}} t]\!] \leq [\![t_n \mathrel{\hat{=}} 0]\!].$$

But $[\![\text{'}t_n \in S^{\#}\setminus\{0\}\text{'}]\!] = 1$, and so $[\![t_n \mathrel{\hat{=}} s]\!] \wedge [\![t_n \mathrel{\hat{=}} t]\!] = 0$. Since $\mathbb{B}$ is ccc, $[\![t_n \mathrel{\hat{=}} s]\!] = 0$ for all but countably many s in S. For $n \in \mathbb{N}$, set $S_n = \{s \in S : [\![t_n \mathrel{\hat{=}} s]\!] \neq 0\}$. Take $s \in S$. Then $s \neq 0$, and so $[\![s \mathrel{\hat{=}} 0]\!] \neq 1$. Thus $[\![s \mathrel{\hat{=}} t_n]\!] \neq 0$ for some $n \in \mathbb{N}$, and so $S = \bigcup\{S_n : n \in \mathbb{N}\}$, whence S is a countable set. Thus $\mathfrak{C}^*$ is ccc.

By hypothesis, $[\![\text{'}|v| \leqslant \mathfrak{c}\text{'}]\!] = 1$. Since $|B| = \mathfrak{c}$, it follows from 7.28 that $[\![\text{'}\mathfrak{c} \leqslant \check{\mathfrak{c}}\text{'}]\!] = 1$. Thus, by the fullness lemma, there exists $t \in V^{\mathbb{B}}$ with

$$[\![\text{'}t \text{ is a surjection from } \check{\mathfrak{c}} \text{ onto } v\text{'}]\!] = 1,$$

and, for each $s \in C^*$, there exists $t_s \in V^{\mathbb{B}}$ with

$$[\![\text{'}t_s \text{ is the least ordinal at which } t \text{ takes the value } s\text{'}]\!] = 1.$$

For $s \in C^*$, define

$$f_s : \alpha \mapsto [\![\check{\alpha} \mathrel{\hat{=}} t_s]\!], \quad \mathfrak{c} \to B.$$

If $\alpha \neq \beta$, then $[\![\check{\alpha} \mathrel{\hat{=}} t_s]\!] \wedge [\![\check{\beta} \mathrel{\hat{=}} t_s]\!] = 0$, and so, since $\mathbb{B}$ is ccc, $\{\alpha : f_s(\alpha) \neq 0\}$ is countable. Since $[\![s \mathrel{\hat{\in}} v]\!] = 1$, $\bigvee\{[\![\check{\beta} \mathrel{\hat{=}} t_s]\!] : \beta < \mathfrak{c}\} = 1$. This implies that, if $s, t \in C^*$ with $f_s = f_t$, then $s \sim t$, and hence, by (a), that $s = t$. Thus the map $s \mapsto f_s$ is an injection.

It follows that

$$|C^*| \leqslant |\{f \in B^{\mathfrak{c}} : |\{\alpha : f(\alpha) \neq 0\}| \leqslant \aleph_0\}|$$

$$= |B|^{\aleph_0} = \mathfrak{c}^{\aleph_0} = \mathfrak{c}.$$

Since $|C^*| \geqslant |B^*| = |B|$, $|C^*| = \mathfrak{c}$, as required.

This completes the proof of the theorem. ∎

We now fix a complete Boolean algebra $\mathbb{B}$, a complete Boolean algebra $\mathbb{C}$ containing $\mathbb{B}$ as a complete subalgebra, and a term $w \in V^{\mathbb{B}}$ such that

$$[\![\text{'} w = \check{\mathbb{C}}/\sim_G \text{'}]\!]^{\mathbb{B}} = 1$$

(so that, by 8.7, $[\![\text{'} w$ is a complete Boolean algebra' $]\!]^{\mathbb{B}} = 1$).

The next step is to define a class $V^{(w)}$ by setting

$$V^{(w)} = \{t \in V^{\mathbb{B}} : [\![\text{'} t \in V^{w} \text{'}]\!]^{\mathbb{B}} = 1\}.$$

(To be precise, $[\![\text{'} t \in V^{w} \text{'}]\!]^{\mathbb{B}} = [\![\psi_o[t,w]]\!]^{\mathbb{B}}$, where $\psi_o(x,y)$ is the formula which formalizes 'y is a complete Boolean algebra and x belongs to V^{y}'.)

Our proof of the iteration theorem involves a map e from $V^{\mathbb{C}}$ to $V^{(w)}$. We shall give a quite explicit definition of e; texts in logic avoid, perhaps wisely, the detailed verifications that we give, for they infer the existence of e from general considerations. In fact, we define $e|V_\alpha^{\mathbb{C}}$ by induction on the ordinal α.

First set $e(\emptyset) = \emptyset$.

Now suppose that $e|V_\alpha^{\mathbb{C}}$ has been defined, and take $t \in V_{\alpha+1}^{\mathbb{C}} \setminus V_\alpha^{\mathbb{C}}$. Define f_t by

$$\left.\begin{aligned} &\text{dom } f_t = \text{dom } t, \text{ and, for } s \in \text{dom } f_t, \\ &f_t(s) = \bigvee\{[\![e(s) \mathrel{\hat{=}} e(r)]\!]^{\mathbb{B}} \wedge [\![r \mathrel{\hat{\in}} t]\!]^{\mathbb{C}} : r \in \text{dom } t\}. \end{aligned}\right\} \quad (5)$$

Then f_t is a function from dom t to C. Certainly

$$[\![s \mathrel{\hat{\in}} t]\!]^{\mathbb{C}} \leq f_t(s) \quad (s \in \text{dom } t). \quad (6)$$

For $s_1, s_2 \in \text{dom } t$, we have

$$[\![e(s_1) \mathrel{\hat{=}} e(s_2)]\!]^{\mathbb{B}} \leq [\![[f_t(s_1)]_G \mathrel{\hat{=}} [f_t(s_2)]_G]\!]^{\mathbb{B}}: \quad (7)$$

to see this, it is sufficient by 8.6(ii) to check that

$$[\![e(s_1) \mathrel{\hat{=}} e(s_2)]\!]^{\mathbf{B}} \wedge f_t(s_1) =$$

$$[\![e(s_1) \mathrel{\hat{=}} e(s_2)]\!]^{\mathbf{B}} \wedge f_t(s_2),$$

and this is immediate from the definition (5). (The reason for defining f_t is that the analogue of (7) for t may fail.) Now set

$$e(t) = \{\{\{e(s)\}^{\#}, \{e(s), [f_t(s)]_G\}^{\#}\}^{\#} : s \in \text{dom } t\}^{\#}$$

(We are determined to be quite explicit!) Thus we have defined $e(t)$ so that

$$[\![\text{'}e(t) \text{ is a function with domain } \{e(s) : s \in \text{dom } t\}^{\#}\text{'}]\!]^{\mathbf{B}} = 1, \tag{8}$$

and so that, for each $s \in \text{dom } t$,

$$[\![\text{'the value of } e(t) \text{ at } e(s) \text{ is } [f_t(s)]_G\text{'}]\!]^{\mathbf{B}} = 1. \tag{9}$$

It follows that $[\![\text{'}e(t) \in V^{\omega}\text{'}]\!]^{\mathbf{B}} = 1$, and so $e(t) \in V^{(\omega)}$.

The inductive definition of e continues.

In summary, for each term t in $V^{\mathbf{C}}$, $e(t)$ is a term in $V^{\mathbf{B}}$ which is also a term for a "term in V^{ω}".

8.9 LEMMA

(i) $[\![e(s) \mathrel{\hat{=}} e(t)]\!]^{\mathbf{B}} \leqslant [\![s \mathrel{\hat{=}} t]\!]^{\mathbf{C}}$ $\quad (s,t \in V^{\mathbf{C}})$.

(ii) For $s \in V^{\mathbf{C}}$, $f_s(p) = [\![p \mathrel{\hat{\in}} s]\!]^{\mathbf{C}}$ $\quad (p \in \text{dom } s)$.

Proof

The proof is by induction on the position of $(\rho(s),\rho(t))$ in $(\underset{\sim}{\text{Ord}}^2,<)$, where ρ denotes rank in $V^{\mathbf{C}}$. Statement (i) holds if either $s = \emptyset$ or $t = \emptyset$

because $[\![e(s) \mathrel{\hat{=}} \emptyset]\!]^{\mathbf{B}} > 0$ if and only if $s \neq \emptyset$, and (ii) is vacuous if $t = \emptyset$.

Now suppose that $\rho(s), \rho(t) > 0$ and that (i) holds for each $p, q \in V^{\mathbf{B}}$ with $(\rho(p), \rho(q)) < (\rho(s), \rho(t))$. Take $p \in \operatorname{dom} s$. Then

$$\begin{aligned} f_s(p) &= \bigvee\{[\![e(p) \mathrel{\hat{=}} e(r)]\!]^{\mathbf{B}} \wedge [\![r \mathrel{\hat{\in}} s]\!]^{\mathbf{C}} : r \in \operatorname{dom} s\} \\ &\leq \bigvee\{[\![p \mathrel{\hat{=}} r]\!]^{\mathbf{C}} \wedge [\![r \mathrel{\hat{\in}} s]\!]^{\mathbf{C}} : r \in \operatorname{dom} s\} \\ &\qquad\qquad\qquad\qquad \text{(inductive hypothesis)} \\ &\leq [\![p \mathrel{\hat{\in}} s]\!]^{\mathbf{C}}, \end{aligned}$$

and so, using (6), $f_s(p) = [\![p \mathrel{\hat{\in}} s]\!]^{\mathbf{C}}$. Thus (ii) holds for s. A similar argument shows that (ii) holds for t.

Write $b = [\![e(s) \mathrel{\hat{=}} e(t)]\!]^{\mathbf{B}}$, and take $p \in \operatorname{dom} s$ and $q \in \operatorname{dom} t$. Then, by (8) and (9),

$$b \wedge [\![e(p) \mathrel{\hat{=}} e(q)]\!]^{\mathbf{B}} \leq [\![[f_s(p)]_G \mathrel{\hat{=}} [f_t(q)]_G]\!]^{\mathbf{B}},$$

and so, by 8.6(iv),

$$\begin{aligned} & b \wedge [\![e(p) \mathrel{\hat{=}} e(q)]\!]^{\mathbf{B}} \wedge f_s(p) \\ &\qquad \leq b \wedge [\![e(p) \mathrel{\hat{=}} e(q)]\!]^{\mathbf{B}} \wedge f_t(q). \end{aligned}$$

Thus, since (ii) holds for s and t,

$$\begin{aligned} & b \wedge [\![e(p) \mathrel{\hat{=}} e(q)]\!]^{\mathbf{B}} \wedge [\![p \mathrel{\hat{\in}} s]\!]^{\mathbf{C}} \\ &\quad \leq b \wedge [\![e(p) \mathrel{\hat{=}} e(q)]\!]^{\mathbf{B}} \wedge [\![q \mathrel{\hat{\in}} t]\!]^{\mathbf{C}} \\ &\quad \leq b \wedge [\![p \mathrel{\hat{=}} q]\!]^{\mathbf{C}} \wedge [\![q \mathrel{\hat{\in}} t]\!]^{\mathbf{C}} \\ &\qquad\qquad\qquad\qquad \text{(inductive hypothesis)} \\ &\quad \leq b \wedge [\![p \mathrel{\hat{\in}} t]\!]^{\mathbf{C}}. \end{aligned}$$

Since $[\![$ 'e(p) belongs to the domain of e(s)' $]\!] = 1$, we have

$$b \leqslant [\![\text{ 'e}(p) \text{ belongs to the domain of } e(t)\text{' }]\!]^{\mathbb{B}}$$

$$= \bigvee\{ [\![e(p) \mathrel{\hat{=}} e(q)]\!]^{\mathbb{B}} : q \in \text{dom } t\},$$

and so $b \wedge [\![p \mathrel{\hat{\in}} s]\!]^{\mathbb{C}} \leqslant b \wedge [\![p \mathrel{\hat{\in}} t]\!]^{\mathbb{C}}$. This holds for each $p \in \text{dom } s$, and so $b \leqslant [\![s \subset t]\!]^{\mathbb{C}}$. Similarly, we have $b \leqslant [\![t \subset s]\!]^{\mathbb{C}}$, and so (i) follows.

The induction continues. ∎

8.10 LEMMA

Let $s,t \in V^{\mathbb{C}}$. Then:

(i) $[\![s \mathrel{\hat{\in}} t]\!]^{\mathbb{C}} = c$ if and only if

$$[\![\text{ '}[\![e(s) \mathrel{\hat{\in}} e(t)]\!]^{\omega} = [c]_G\text{' }]\!]^{\mathbb{B}} = 1;$$

(ii) $[\![s \mathrel{\hat{=}} t]\!]^{\mathbb{C}} = c$ if and only if

$$[\![\text{ '}[\![e(s) \mathrel{\hat{=}} e(t)]\!]^{\omega} = [c]_G\text{' }]\!]^{\mathbb{B}} = 1.$$

<u>Proof</u>

The proof is by induction on the position of $(\rho(s),\rho(t))$ in $(\text{Ord}^2,<)$. Suppose then that the statements hold for $p,q \in V^{\mathbb{C}}$ with $(\rho(p),\rho(q)) < (\rho(s),\rho(t))$ in $(\text{Ord}^2,<)$.

If $t = \emptyset$, then (i) is immediate.

Now suppose that $\rho(t) > 0$. Take $c \in C$ such that

$$[\![\text{ '}[\![e(s) \mathrel{\hat{\in}} e(t)]\!] = [c]_G\text{' }]\!]^{\mathbb{B}} = 1,$$

and, for $p \in \text{dom } t$, take $c_p \in C$ such that

$$[\![\text{ '}[\![e(s) \mathrel{\hat{=}} e(p)]\!]^{\omega} \wedge [f_t(p)]_G = [c_p]_G\text{' }]\!]^{\mathbb{B}} = 1.$$

Set $T = \{c_p : p \in \text{dom } t\}$, and set $S = \{[c_p]_G : p \in \text{dom } t\}^{\#}$.

Then $[\!['VS = [VT]_G \text{ in } w']\!]^{B} = [\!['VS = [c]_G \text{ in } w']\!]^{B} = 1$, and so

$$c = VT = V\{c_p : p \in \text{dom } t\}. \tag{10}$$

By the inductive hypothesis,

$$[\![' [\![e(s) \hat{=} e(p)]\!]^{w} = [[\![s \hat{=} p]\!]^{C}]_G ']\!]^{B} = 1$$

for $p \in \text{dom } t$, and so, by 8.6(i) and (iii), we have $[\![s \hat{=} p]\!]^{C} \wedge f_t(p) = c_p$. By 8.9(ii), $f_t(p) = [\![p \hat{\in} t]\!]^{C}$, and so, from (10), $c = [\![s \hat{\in} t]\!]^{C}$, giving (i).

Next, one shows that $[\![s \subset t]\!]^{C} = c$ if and only if

$$[\![' [\![e(s) \subset e(t)]\!]^{w} = [c]_G ']\!]^{B} = 1.$$

We omit the details; they are very similar to those just given.

By permuting s and t, we obtain the second half of (ii): for this, the induction hypothesis is applied to pairs (r,s) where $r \in \text{dom } t$, which is allowed.

The induction continues. ∎

We now show that e is "onto $V^{(w)}$ up to equivalence". It is convenient to make the following definitions: for $\alpha \in \underset{\sim}{\text{Ord}}$,

$$V_{\alpha}^{(w)} = \{t \in V^{(w)} : [\!['\rho(t) < \check{\alpha} \text{ in } V^{w}']\!]^{B} = 1\}.$$

It is easy to see by induction on rank that, if $t \in V_{\alpha}^{C}$, then $e(t) \in V_{\alpha}^{(w)}$.

8.11 LEMMA

Let $s \in V_{\alpha}^{(w)}$. Then there exists $t \in V_{\alpha+1}^{C}$ such that $[\!['e(t) \sim s \text{ in } V^{w}']\!]^{B} = 1$.

<u>Proof</u>

The proof is by induction on α. Since $e(\emptyset) = \emptyset$ and $\{\emptyset\} = V_1^{(w)} \subset V_2^{(w)}$, the result holds for $\alpha = 1$.

Now suppose that $\alpha > 1$ and that the result holds for each β with $\beta < \alpha$. For $s \in V_\alpha^{(w)}$, define $t \in V_{\alpha+1}^{\mathbf{C}}$ by

$$\text{dom } t = V_\alpha^{\mathbf{C}}, \text{ and, for } p \in \text{dom } t,$$

$$t(p) = c \text{ if } [\![\text{'}[\![e(p) \mathrel{\hat{\in}} s]\!]^w = [c]_G\text{'}]\!]^{\mathbf{B}} = 1.$$

For $p, q \in \text{dom } t$,

$$[\![\text{'}[\![e(p) \mathrel{\hat{=}} e(q)]\!]^w \wedge [\![e(q) \mathrel{\hat{\in}} s]\!]^w \leq [\![e(p) \mathrel{\hat{\in}} s]\!]^{w}\text{'}]\!]^{\mathbf{B}} = 1,$$

and so, by 8.10(ii),

$$[\![\text{'}[[\![p \mathrel{\hat{=}} q]\!]^{\mathbf{C}}]_G \wedge [t(q)]_G \leq [t(p)]_G\text{'}]\!]^{\mathbf{B}} = 1.$$

Hence $[\![p \mathrel{\hat{=}} q]\!]^{\mathbf{C}} \wedge t(q) \leq t(p)$, and so

$$[\![p \mathrel{\hat{\in}} t]\!]^{\mathbf{C}} = \bigvee\{t(q) \wedge [\![p \mathrel{\hat{=}} q]\!]^{\mathbf{C}} : q \in \text{dom } t\} = t(p).$$

By 8.9(ii), $t(p) = f_t(p)$ $(p \in \text{dom } t)$.

It follows from (8) and (9) that

$$[\![\text{'the value of } e(t) \text{ at } e(p) \text{ is } [\![e(p) \mathrel{\hat{\in}} s]\!]^{w}\text{'}]\!]^{\mathbf{B}} = 1,$$

and that $[\![\text{'}[\![e(t) \subset s]\!]^w = 1\text{'}]\!]^{\mathbf{B}} = 1.$

To complete the proof, we must show that $[\![\text{'}[\![s \subset e(t)]\!]^w = 1\text{'}]\!]^{\mathbf{B}} = 1$. Suppose that this is false. Then there exists $r \in V^{\mathbf{B}}$ such that $b > 0$, where

$$b = [\![\text{'}r \in \text{dom } s\text{'}]\!]^{\mathbf{B}} \wedge [\![\text{'}[\![r \mathrel{\hat{\in}} s]\!]^w \wedge [\![r \mathrel{\hat{\notin}} e(t)]\!]^w > 0\text{'}]\!]^{\mathbf{B}}.$$

Since $s \in V_\alpha^{(w)}$, we have $b \leqslant [\!['r \in \mathrm{dom}\ s \wedge \rho(s) < \check{\alpha}']\!]^{\mathbf{B}}$, and so there exists $a \in B$ with $0 < a \leqslant b$ and $\beta < \alpha$ such that

$$[\!['\rho(r) < \check{\beta} \text{ in } V^{w}']\!]^{\mathbf{B}} = a.$$

Choose $q \in V^{\mathbf{B}}$ with $[\![q \mathrel{\hat{=}} r]\!]^{\mathbf{B}} \geqslant a$ and $[\![q \mathrel{\hat{=}} \emptyset]\!]^{\mathbf{B}} \geqslant a'$. Then $[\!['\rho(q) < \check{\beta} \text{ in } V^{w}']\!]^{\mathbf{B}} = 1$, and so, by the inductive hypothesis, there exists $p \in V_{\beta+1}^{\mathbf{C}} \subset V_\alpha^{\mathbf{C}}$ such that $[\!['e(p) \sim q \text{ in } V^{w}']\!]^{\mathbf{B}} = 1$. We have $[\![e(p) \mathrel{\hat{=}} r]\!]^{\mathbf{B}} \geqslant a$, and so

$$[\![' [\![e(p) \mathrel{\hat{\in}} s]\!]^{w} \wedge [\![e(p) \mathrel{\hat{\notin}} e(t)]\!]^{w} > 0']\!]^{\mathbf{B}} \geqslant a. \qquad (11)$$

By definition, $\mathrm{dom}\ t = V_\alpha^{\mathbf{C}}$, and so $p \in \mathrm{dom}\ t$. It follows from (9) and the fact that $t(p) = f_t(p)$ that

$$[\![' [\![e(p) \mathrel{\hat{\in}} e(t)]\!]^{w} \geqslant [t(p)]_G ']\!]^{\mathbf{B}} = 1.$$

Thus

$$[\![' [\![e(p) \mathrel{\hat{\in}} e(t)]\!]^{w} \geqslant [\![e(p) \mathrel{\hat{\in}} s]\!]^{w} ']\!]^{\mathbf{B}} = 1.$$

But this is a contradiction of (11).

The proof of the lemma is complete. ∎

Let $\phi(x_1, \ldots, x_n)$ be a formula.

8.12 THEOREM

Let $\mathbf{B}$ be a complete Boolean algebra, let $\mathbf{C}$ be a complete Boolean algebra containing $\mathbf{B}$ as a complete subalgebra and let $w \in V^{\mathbf{B}}$ be such that $[\!['w = \check{\mathbf{C}}/\sim_G ']\!] = 1$. Then, for $t_1, \ldots, t_n \in V^{\mathbf{C}}$ and $c \in C$,

$$[\![\phi[t_1, \ldots, t_n]]\!]^{\mathbf{C}} = c \text{ if and only if}$$
$$[\![' [\![\phi[e(t_1), \ldots, e(t_n)]]\!]^{w} = [c]_G ']\!]^{\mathbf{B}} = 1.$$

<u>Proof</u>

Let Φ be the set of subformulae ψ of ϕ such

that the result holds for ψ.

By 8.10, Φ contains all the atomic formulae in ϕ. By 8.6(i), Φ contains $\psi \vee \chi$ whenever Φ contains ψ and χ and $\psi \vee \chi$ is a subformula of ϕ. Similarly, Φ contains $\neg\psi$ whenever Φ contains ψ and $\neg\psi$ is a subformula.

Suppose that Φ contains $\psi(x_1)$ and that $\exists x_1\psi$ is a subformula. Set $c = [\![\exists x_1\psi]\!]^{\mathbf{C}}$, and take $d \in C$ with $[\![\text{'}[\![\exists x_1\psi]\!]^{\omega} = [d]_G\text{'}]\!]^{\mathbf{B}} = 1$. By the fullness lemma, there exists $t \in V^{\mathbf{C}}$ with $[\![\psi[t]]\!]^{\mathbf{C}} = c$. Since $[\![\text{'}[\![\psi[e(t)]]\!]^{\omega} = [c]_G\text{'}]\!]^{\mathbf{B}} = 1$, it follows that $c \leqslant d$. By the fullness lemma again,

$$[\![\text{'there exists } t \in V^{\omega} \text{ with } [\![\psi[t]]\!]^{\omega} = [d]_G\text{'}]\!]^{\mathbf{B}} = 1,$$

and so, by a third application of the fullness lemma, there exists $s \in V^{\mathbf{B}}$ with

$$[\![\text{'}[\![\psi[s]]\!]^{\omega} = [d]_G\text{'}]\!]^{\mathbf{B}} = [\![\text{'}s \in V^{\omega}\text{'}]\!]^{\mathbf{B}} = 1.$$

Now $s \in V^{(\omega)}$, and so it follows from 8.11 that there exists $u \in V^{\mathbf{C}}$ with $[\![\text{'}e(u) \sim s \text{ in } V^{\omega}\text{'}]\!]^{\mathbf{B}} = 1$. Since $\psi \in \Phi$, $[\![\psi[u]]\!]^{\mathbf{C}} = d$, and so $d \leqslant c$. Thus $c = d$, and hence $\exists x_1\psi \in \Phi$.

It follows that all subformulae of ϕ belong to Φ, and in particular the result holds for ϕ itself. This completes the proof of the theorem. ∎

We need one last lemma before we can complete our analysis of $V^{(\omega)}$ in terms of $V^{\mathbf{C}}$.

Let $u \in V^{\mathbf{B}}$ with

$$[\![\text{'}u \text{ is a complete Boolean algebra'}]\!]^{\mathbf{B}} = 1.$$

For $t \in V^{\mathbf{B}}$, we have $\check{t} \in V^{\mathbf{B}}$. But inside $V^{\mathbf{B}}$ we can form "$\check{t}$ relative to u". In an attempt to avoid confusion, we denote this latter term by $(t)^{\vee}_{u}$. Of course, $(t)^{\vee}_{u}$ is only defined up to equivalence in $V^{\mathbf{B}}$, but this will not matter because $(t)^{\vee}_{u}$ will only appear in expressions of the form $[\![\text{'}\ldots\text{'}]\!]^{B}$.

8.13 LEMMA

Let $t \in V^{B}$. Then $[\![\text{'}e(t) \sim (t)^{\vee}_{w} \text{ in } V^{w}\text{'}]\!] = 1.$

<u>Proof</u>

Let $s \in \text{dom } t$. The main point of the proof is that the following equation holds:

$$[\![\text{'}[\![e(s)\ \hat{\in}\ e(t)]\!]^{w} = [\![(s)^{\vee}_{w}\ \hat{\in}\ (t)^{\vee}_{w}]\!]^{w}\text{'}]\!] = 1. \quad (12)$$

To see this, set $b = [\![s\ \hat{\in}\ t]\!]^{B}$. By 8.10(i),

$$[\![\text{'}[\![e(s)\ \hat{\in}\ e(t)]\!]^{w} = [b]_{G}\text{'}]\!]^{B} = 1,$$

and so

$$[\![\text{'}[\![e(s)\ \hat{\in}\ e(t)]\!]^{w} = 1\text{'}]\!]^{B} = b,$$

$$[\![\text{'}[\![e(s)\ \hat{\in}\ e(t)]\!]^{w} = 0\text{'}]\!]^{B} = b'.$$

Also, we have

$$[\![\text{'}[\![(s)^{\vee}_{w}\ \hat{\in}\ (t)^{\vee}_{w}]\!]^{w} = 1\text{'}]\!]^{B} = [\![s\ \hat{\in}\ t]\!]^{B} = b,$$

$$[\![\text{'}[\![(s)^{\vee}_{w}\ \hat{\in}\ (t)^{\vee}_{w}]\!]^{w} = 0\text{'}]\!]^{B} = [\![s\ \hat{\notin}\ t]\!]^{B} = b',$$

and so (12) follows.

We now prove the lemma by induction on $\rho(t)$. The result is trivial if $\rho(t) = 0$. Suppose that $\rho(t) > 0$ and that the result holds for each $s \in V^{B}$ with $\rho(s) < \rho(t)$.

Take $s \in \text{dom } t$. By the inductive hypothesis and (12),

$$[\![\text{'}[\![e(s)\ \hat{\in}\ e(t)]\!]^{w} = [\![e(s)\ \hat{\in}\ (t)^{\vee}_{w}]\!]^{w}\text{'}]\!]^{B} = 1,$$

and so, taking account of (12), we have

$$[\![\text{'} [\![e(t) \subset (t)^{\vee}_{w}]\!]^{w} = 1\text{'}]\!]^{\mathcal{B}} = 1.$$

Also by the inductive hypothesis and (12),

$$[\![\text{'} [\![(s)^{\vee}_{w} \hat{\in} (t)^{\vee}_{w}]\!]^{w} = [\![(s)^{\vee}_{w} \hat{\in} e(t)]\!]^{w} \text{'}]\!]^{\mathcal{B}} = 1,$$

and so, since $[\![t \subset (\text{dom } t)^{\#}]\!]^{\mathcal{B}} = 1$, we have

$$[\![\text{'} [\![(t)^{\vee}_{w} \subset e(t)]\!]^{w} = 1\text{'}]\!]^{\mathcal{B}} = 1.$$

The result for t follows, and the induction continues. ∎

Let $\phi(x_1,\ldots,x_n)$ be a formula.

We need an observation: if $\mathcal{B}_1$ and $\mathcal{B}_2$ are complete Boolean algebras, if $\pi: \mathcal{B}_1 \to \mathcal{B}_2$ is a Boolean isomorphism, and if $a_1,\ldots,a_n \in V$, then

$$\pi([\![\phi[\check{a}_1,\ldots,\check{a}_n]]\!]^{\mathcal{B}_1}) = [\![\phi[\check{a}_1,\ldots,\check{a}_n]]\!]^{\mathcal{B}_2}.$$

8.14 THEOREM *(Iteration theorem)*

(i) Let $\mathcal{B}$ be a complete Boolean algebra, and let $t_1,\ldots,t_n$ and u be terms in $V^{\mathcal{B}}$ such that

$$[\![\text{'}u \text{ is a complete Boolean algebra'}]\!]^{\mathcal{B}}$$

$$= [\![\text{'} [\![\phi[(t_1)^{\vee}_{u},\ldots,(t_n)^{\vee}_{u}]]\!]^{u} = 1\text{'}]\!]^{\mathcal{B}} = 1.$$

Then there is a complete Boolean algebra $\mathcal{C}$ containing $\mathcal{B}$ as a complete subalgebra such that $[\![\phi[t_1,\ldots,t_n]]\!]^{\mathcal{C}} = 1$.

(ii) Suppose, further, that $\mathcal{B}$ is ccc and has cardinality $2^{\aleph_0}$ and that

$$[\![\text{'}u \text{ is ccc and has cardinality } \leqslant 2^{\aleph_0}\text{'}]\!]^{\mathcal{B}} = 1.$$

Then we can suppose that $\mathcal{C}$ is ccc and has cardinality $2^{\aleph_0}$.

Proof

By 8.8, there is a complete Boolean algebra $\mathbb{C}$ containing $\mathbb{B}$ as a complete subalgebra such that

$$[\!['u \text{ is isomorphic to } \check{\mathbb{C}}/\sim_G']\!]^{\mathbb{B}} = 1.$$

Choose $w \in V^{\mathbb{B}}$ with $[\!['w = \check{\mathbb{C}}/\sim_G']\!]^{\mathbb{B}} = 1$. Since $[\!['u \text{ is isomorphic to } w']\!]^{\mathbb{B}} = 1$, the above observation shows that

$$[\![' [\![\phi[(t_1)^{\vee}_{w},\ldots,(t_n)^{\vee}_{w}]]\!]^{w} = 1']\!]^{\mathbb{B}} = 1.$$

By 8.13, $[\!['\phi[e(t_1),\ldots,e(t_n)]]\!]^{u} = 1']\!]^{\mathbb{B}} = 1$, and so, by Theorem 8.12, $[\![\phi[t_1,\ldots,t_n]]\!]^{\mathbb{C}} = 1$.

(ii) This follows from 8.8(ii). ∎

Our strategy for building the complete Boolean algebra $\mathbb{A}$ for which $[\![\mathrm{MA}]\!]^{\mathbb{A}} = 1$ is to express $\mathbb{A}$ as a union of complete subalgebras which are themselves built by successively eliminating potential counter-examples. To carry out this strategy we shall require the following result.

8.15 PROPOSITION

Let ξ be an ordinal and let $\langle \mathbb{B}_\alpha : \alpha < \xi \rangle$ be a sequence of complete ccc Boolean algebras such that:

(a) $\mathbb{B}_\alpha$ is a complete subalgebra of $\mathbb{B}_\beta$ whenever $\alpha < \beta < \xi$;

(b) $\bigcup\{B_\beta \setminus \{0\} : \beta < \alpha\}$ is dense in $\mathbb{B}_\alpha$ whenever α is a limit ordinal with $\alpha < \xi$.

Let $\mathbb{B} = \bigcup_{\alpha<\xi} \mathbb{B}_\alpha$. Then $\mathbb{B}$ is a ccc Boolean algebra, and $\mathbb{B}$ is complete in the case where $\mathrm{cf}\,\xi \geq \omega_1$.

Proof

First suppose that $\mathrm{cf}\,\xi = \omega$, say $\langle \alpha_n : n \in \omega \rangle$ is cofinal in ξ. Let A be an antichain in $\mathbb{B}$. Then $A \cap B_{\alpha_n}$ is an antichain in $\mathbb{B}_{\alpha_n}$, and so is countable. Since

$A = \bigcup(A \cap B_{\alpha_n})$, A is countable, and so $\mathcal{B}$ is ccc.

Secondly, suppose that $\mathrm{cf}\ \xi > \omega_1$. Suppose that A is an uncountable antichain in $\mathcal{B}$. Then A contains a subset C of cardinality $\aleph_1$. But $C \subset B_\alpha$ for some $\alpha < \xi$, a contradiction. Thus $\mathcal{B}$ is ccc.

We now come to the main case: suppose that $\mathrm{cf}\ \xi = \omega_1$. By passing to a subsequence, we may suppose that $\xi = \omega_1$.

For $b \in B$ and $\alpha < \omega_1$, set

$$\pi_\alpha(b) = \bigwedge\{c \in B_\alpha : b \leqslant c\}.$$

Then $b \leqslant \pi_\alpha(b)$, $\pi_\alpha(b) \in B_\alpha$ because $\mathcal{B}_\alpha$ is complete, and $\pi_\alpha(b) \geqslant \pi_\beta(b)$ whenever $\alpha < \beta < \omega_1$ because $\mathcal{B}_\alpha$ is a complete subalgebra of $\mathcal{B}_\beta$.

Let A be an antichain in $\mathcal{B}$. For $\alpha < \beta < \omega_1$, set $A_{\alpha,\beta} = A \cap (B_\beta \backslash B_\alpha)$, and, for $\alpha < \omega_1$, set $A_\alpha = A \setminus (A \cap B_\alpha) = \bigcup\{A_{\alpha,\beta} : \alpha < \beta\}$. We *claim* that, for each $\alpha < \omega_1$, there exists $\beta \in (\alpha, \omega_1)$ with

$$\bigvee\pi_\alpha(A_\alpha) = \bigvee\pi_\alpha(A_{\alpha,\beta}).$$

For set $d_\beta = \bigvee\pi_\alpha(A_{\alpha,\beta})$. Then $\langle d_\beta : \alpha < \beta < \omega_1\rangle$ is an increasing sequence in the ccc Boolean algebra $\mathcal{B}_\alpha$, and so there exists $d \in B_\alpha$ such that $d_\alpha = d$ eventually. Clearly $\bigvee\pi_\alpha(A_\alpha) = d$, and this establishes the claim.

By 5.7, there is a limit ordinal $\eta < \omega_1$ such that, for each $\alpha < \eta$, there exists $\beta \in (\alpha, \eta)$ with $\bigvee\pi_\alpha(A_\alpha) = \bigvee\pi_\alpha(A_{\alpha,\beta})$. We now *claim* that $A \subset B_\eta$. For suppose not, and take $a \in A\backslash B_\eta$. Then $\pi_\eta(a) \in B_\eta\backslash\{0\}$, and so, by hypothesis (b), there exists $\alpha < \eta$ and $c \in B_\alpha\backslash\{0\}$ with $c \leqslant \pi_\eta(a)$. We have $\bigvee\pi_\alpha(A_\alpha) = \bigvee\pi_\alpha(A_{\alpha,\eta})$. For each $d \in A_{\alpha,\eta}$, $a \leqslant d'$, and so $\pi_\eta(a) \leqslant d'$. Thus $c' \in B_\alpha$ and $d \leqslant c'$, and so $\pi_\alpha(d) \leqslant c'$, i.e., $c \wedge \pi_\alpha(d) = 0$. Hence $c \wedge (\bigvee\pi_\alpha(A_\alpha)) = 0$. But $a \in A_\alpha$, and $\pi_\alpha(a) \geqslant \pi_\eta(a) \geqslant c$, and so we have a contradiction. Thus $A \subset B_\eta$. Since $\mathcal{B}_\eta$ is ccc, A is countable, and so $\mathcal{B}$ is ccc.

Now suppose that $\mathrm{cf}\,\xi \geqslant \omega_1$, and let S be a subset of B. Set $d_\alpha = \bigvee(S \cap B_\alpha)$. Then $\langle d_\alpha : \alpha < \xi \rangle$ is an increasing sequence in $\mathcal{B}$, and it eventually takes the constant value d, say, because $\mathcal{B}$ is ccc. Clearly $d = \bigvee S$ in $\mathcal{B}$, and so $\mathcal{B}$ is complete. ∎

In Definition 8.2, we defined Δ_o-formulae, and in Proposition 8.3 we showed that, if $\phi(x_1,\ldots,x_n)$ is a Δ_o-formula and if $\mathcal{B}$ is a complete subalgebra of a complete Boolean algebra $\mathcal{C}$, then $[\![\phi[t_1,\ldots,t_n]]\!]^{\mathcal{B}} = [\![\phi[t_1,\ldots,t_n]]\!]^{\mathcal{C}}$ whenever $t_1,\ldots,t_n \in V^{\mathcal{B}}$. We now list some typical properties which can be specified by Δ_o-formulae and to which this result will be applied. The verifications are straightforward and tedious, and will be omitted.

(i) 'x is a function from y to z'

(ii) 'x is a bijection from y to z'

(iii) 'x is a partially ordered set, and y is an antichain in x'

(iv) 'x is a partially ordered set, y is a family of dense sets in x, z is a y-generic filter in x'

(v) 'x and y are partially ordered sets, and $z : x \to y$ is an embedding'

(vi) 'x and y are totally ordered sets, and $z : x \to y$ is an embedding whose range is cofinal in y'

(vii) 'x and y belong to $\mathbb{N}^{\mathbb{N}}$ and $x <_F y$'

(viii) 'x is a pregap in $(\mathbb{N}^{\mathbb{N}}, <_F)$'

(ix) 'x and y are equivalent pregaps in $(\mathbb{N}^{\mathbb{N}}, <_F)$'

We have explained that our strategy for building a complete Boolean algebra $\mathcal{A}$ for which $[\![MA]\!]^{\mathcal{A}} = 1$ is to successively eliminate counter-examples to MA. At the same time, we shall build the algebra $\mathcal{A}$ so that $[\![NDH]\!]^{\mathcal{A}} = 1$: our preliminary results in Chapter 6 show that, to do this, it is sufficient to eliminate potential embeddings from $(\mathbb{R},<)$ into $(\mathbb{N}^{\mathbb{N}},<_F)$, and that this follows if we ensure that each embedding of $(\mathbb{Q},<)$ into $(\mathbb{N}^{\mathbb{N}},<_F)$ has a Hausdorff gap in its range, and so cannot be extended to an embedding of $(\mathbb{R},<)$. (We realize that to use fraktur letters for the underlying sets of the partially ordered sets $(\mathbb{Q},<)$ and $(\mathbb{R},<)$ conflicts with our present notational convention, but our typographic possibilities are finite!)

In the preface, we mentioned that the style of analysis that we must undertake in the present chapter requires a certain precision unusual in other areas of mathematics. The next definition, which is a slight reformulation of part of Definition 6.4, exemplifies this. We now say that $r = \langle A,B\rangle$ is an $(\aleph_1,\aleph_1)$-pregap in $(\mathbb{N}^{\mathbb{N}},<_F)$ if $A \cup B$ is totally ordered, if $A < B$, if there is an isotonic map $\pi_1\colon (\omega_1,\in) \to (\mathbb{N}^{\mathbb{N}},<_F)$ such that $\operatorname{ran}\pi_1$ and A are mutually cofinal, and if there is an anti-isotonic map $\pi_2\colon (\omega_1,\in) \to (\mathbb{N}^{\mathbb{N}},<_F)$ such that $\operatorname{ran}\pi_2$ and B are mutually coinitial. The point of this form of the definition is that we can see that each of the stated properties of r, π_1, and π_2, save for those giving the cardinality of the domains of π_1 and π_2, can be given by a Δ_o-formula. For example, to say that $\operatorname{dom}\pi_1$ is an ordinal and that $\operatorname{ran}\pi_1$ and A are mutually cofinal in $(\mathbb{N}^{\mathbb{N}},<_F)$ is a Δ_o-property of r and π_1.

Now properties specified by Δ_o-formulae are preserved in the passage from V to $V^{\mathbb{B}}$, but the cardinal $\aleph_1$ may not be preserved. Thus, if r is an $(\aleph_1,\aleph_1)$-pregap in $(\mathbb{N}^{\mathbb{N}},<_F)$, then

$$[\![\text{'}\check{r} \text{ is an } (\aleph_1,\aleph_1)\text{-pregap in } (\mathbb{N}^{\mathbb{N}},<_F)\text{'}]\!]^{\mathbb{B}} = [\![\text{'}\check{\aleph}_1 = \aleph_1\text{'}]\!]^{\mathbb{B}},$$

$$[\![\text{'}\check{r} \text{ is an } (\aleph_o,\aleph_o)\text{-pregap in } (\mathbb{N}^{\mathbb{N}},<_F)\text{'}]\!]^{\mathbb{B}} = [\![\text{'}\check{\aleph}_1 = \aleph_o\text{'}]\!]^{\mathbb{B}}.$$

But now suppose that $\mathbb{B}$ is ccc. Then, by 7.27, cardinals are preserved in $V^{\mathbb{B}}$, and so $[\![\text{'}\check{\aleph}_1 = \aleph_1\text{'}]\!]^{\mathbb{B}} = 1$. More generally, if $\mathbb{B}$ is a complete subalgebra of a complete ccc

Boolean algebra $\mathbb{C}$, then cardinals are preserved in the passage from $V^{\mathbb{B}}$ to $V^{\mathbb{C}}$, and so, for $t \in V^{\mathbb{B}}$,

$$[\![\text{'}t \text{ is an } (\aleph_1,\aleph_1)\text{-pregap in } (\mathbb{N}^{\mathbb{N}},<_F)\text{'}]\!]^{\mathbb{B}} \leq [\![\text{'}\ldots\text{'}]\!]^{\mathbb{C}}. \tag{12}$$

Now let $r = \langle A,B \rangle$ be a Hausdorff gap in $(\mathbb{N}^{\mathbb{N}},<_F)$. Then there exist maps π_1 and π_2 as above with the additional properties that $\pi_1(\alpha) \leq \pi_2(\alpha)$ $(\alpha < \omega_1)$ and that, for each $\alpha,\beta < \omega_1$ with $\alpha \neq \beta$, either $\pi_1(\alpha) \nleq \pi_2(\beta)$ or $\pi_1(\beta) \nleq \pi_2(\alpha)$. These additional properties of π_1 and π_2 can also be expressed by Δ_o-formulae. Thus, if $\mathbb{B}$ is a complete subalgebra of a complete ccc Boolean algebra $\mathbb{C}$, then

$$[\![\text{'}t \text{ is a Hausdorff gap'}]\!]^{\mathbb{B}} \leq [\![\text{'}\ldots\text{'}]\!]^{\mathbb{C}}. \tag{13}$$

Notice that we have "$\leq$" in (12) and (13): this is because there are more terms in $V^{\mathbb{C}}$ than in $V^{\mathbb{B}}$ which are potential witnesses to the fact that t is an $(\aleph_1,\aleph_1)$-pregap or a Hausdorff gap, respectively. In fact, equality holds in (12) because cofinalities are preserved, a result we shall not need, but equality may fail in (13), and it is this failure that we shall exploit in our construction of a complete Boolean algebra $\mathbb{A}$ such that $[\![\text{NDH}]\!]^{\mathbb{A}} = 1$.

We now give some analogous remarks about partially ordered sets. Let $\mathbb{B}$ be a complete subalgebra of a complete Boolean algebra $\mathbb{C}$, and let $a,t \in V^{\mathbb{B}}$. Then

$$[\![\text{'}t \text{ is a partially ordered set'}]\!]^{\mathbb{B}} = [\![\text{'}\ldots\text{'}]\!]^{\mathbb{C}},$$

and

$$[\![\text{'}a \text{ is an antichain in } t\text{'}]\!]^{\mathbb{B}} = [\![\text{'}\ldots\text{'}]\!]^{\mathbb{C}}.$$

Further, if $\mathbb{C}$ is ccc, then

$$[\![\text{'}a \text{ is uncountable'}]\!]^{\mathbb{B}} = [\![\text{'}\ldots\text{'}]\!]^{\mathbb{C}}$$

because cardinals are preserved from $V^{\mathbb{B}}$ to $V^{\mathbb{C}}$, and so

$$[\![\text{'}t \text{ is a ccc partially ordered set}]\!]^{\mathbb{B}} \geq [\![\text{'...'}]\!]^{\mathbb{C}}; \quad (14)$$

we have "$\geq$" in (14) because there are possibly more terms for uncountable antichains in $V^{\mathbb{C}}$ than in $V^{\mathbb{B}}$.

8.16 LEMMA

Let $\mathbb{B}$ be the completion of a partially ordered set $\mathbb{P}$, and let D be a family of dense subsets of $\mathbb{P}$. Then

$$[\![\text{'there is a } \check{D}\text{-generic filter in } \check{\mathbb{P}}\text{'}]\!]^{\mathbb{B}} = 1.$$

Proof

Let $\pi : P \to B\setminus\{0\}$ be the associated order-preserving map. Since $\pi(P)$ is dense in $\mathbb{B}$, $\pi(d)$ is dense in $\mathbb{B}$ for each $d \in D$, and so $\bigvee\{\pi(p) : p \in d\} = 1$.

Define $F \in V^{\mathbb{B}}$ by

$$\operatorname{dom} F = \{\check{p} : p \in P\}, \quad F(\check{p}) = \pi(p) \quad (p \in P).$$

It is easy to check that $[\![\text{'}F \text{ is a filter in } \check{\mathbb{P}}\text{'}]\!]^{\mathbb{B}} = 1$. Take $d \in D$. Then

$$[\![\text{'}F \cap \check{d} \neq \emptyset\text{'}]\!]^{\mathbb{B}} = \bigvee\{\pi(p) : p \in d\} = 1,$$

and this gives the result. ∎

Let π_1 and π_2 be isotonic and anti-isotonic maps, respectively, from $(\omega_1, \in)$ into $(\mathbb{N}^{\mathbb{N}}, <_F)$ such that $\langle \operatorname{ran} \pi_1, \operatorname{ran} \pi_2 \rangle$ is a pregap. Set $f_\alpha = \pi_1(\alpha)$ and $g_\alpha = \pi_2(\alpha)$ for $\alpha < \omega_1$, and set $t = (\pi_1, \pi_2)$. Let $\mathbb{Q}(t)$ be the partially ordered set with the rather complicated definition that was denoted by $Q(\langle f_\alpha, g_\alpha \rangle)$ in Definition 6.11. The underlying set of $\mathbb{Q}(t)$ is $Q(t)$. For $\xi < \omega_1$, set

$$d(t,\xi) = \{q \in Q(t) : \xi \in \operatorname{supp} q\},$$

and

$$D(t) = \{d(t,\xi) : \xi < \omega_1\}.$$

By 6.13, each $d(t,\xi)$ is dense in $\mathbb{Q}(t)$. Further, 6.13 shows that, if there is a $D(t)$-generic filter on $\mathbb{Q}(t)$, then $\langle f_\alpha, g_\alpha : \alpha < \omega_1\rangle$ is a Hausdorff gap.

8.17 LEMMA

Let $s = \langle \operatorname{ran} \pi_1, \operatorname{ran} \pi_2\rangle$ be the pregap in $(\mathbb{N}^{\mathbb{N}}, <_F)$ specified above. Assume that s is actually a gap in $(\mathbb{N}^{\mathbb{N}}, <_F)$. Then

$$[\![\text{'}\check{s} \text{ is a Hausdorff gap in } (\mathbb{N}^{\mathbb{N}}, <_F)\text{'}]\!]^{\mathbb{B}} = 1,$$

where $\mathbb{B}$ denotes the completion of $\mathbb{Q}(t)$.

Proof

Since s is a gap, Theorem 6.12 proves that $\mathbb{Q}(t)$ is ccc. By 5.16, $\mathbb{B}$ is a ccc Boolean algebra, and so

$$[\![\text{'}\check{s} \text{ is an } (\aleph_1, \aleph_1)\text{-pregap in } (\mathbb{N}^{\mathbb{N}}, <_F)\text{'}]\!]^{\mathbb{B}} = 1.$$

By 8.16,

$$[\![\text{'there is a } D(t)^{\vee}\text{-generic filter on } \mathbb{Q}(t)^{\vee}\text{'}]\!]^{\mathbb{B}} = 1. \quad (15)$$

It is clear that $[\![\text{'}[(f,g)]^{\vee} = [(\check{f},\check{g})]\text{'}]\!]^{\mathbb{B}} = 1$ for $f,g \in \mathbb{N}^{\mathbb{N}}$, where we are using the notation of Chapter 6, and so $[\![\text{'}\mathbb{Q}(t)^{\vee} \hat{=} \mathbb{Q}(\check{t})\text{'}]\!]^{\mathbb{B}} = 1$. Also, for each $\xi < \omega_1$,

$$[\![\text{'}d(t,\xi)^{\vee} = \{q \in \mathbb{Q}(\check{t}) : \check{\xi} \in \operatorname{supp} q\}\text{'}]\!]^{\mathbb{B}} = 1,$$

and so $[\![\text{'}d(t,\xi)^{\vee} \hat{=} d(\check{t},\check{\xi})\text{'}]\!]^{\mathbb{B}} = 1$. Hence, we have $[\![\text{'}D(t)^{\vee} \hat{=} D(\check{t})\text{'}]\!]^{\mathbb{B}} = 1$, and so, from (15),

$$[\![\text{'there is a } D(\check{t})\text{-generic filter on } \mathbb{Q}(\check{t})\text{'}]\!]^{\mathbb{B}} = 1.$$

The result follows from 6.13. ∎

8.18 LEMMA (CH)

Let s be an $(\aleph_1,\aleph_1)$-gap in $(\mathbb{N}^{\mathbb{N}},<_F)$. Then there is a complete ccc Boolean algebra $\mathbb{B}$ of cardinality $\aleph_1$ such that

$$[\![\text{'}\check{s} \text{ is a Hausdorff gap'}]\!]^{\mathbb{B}} = 1.$$

<u>Proof</u>

Choose maps π_1 and π_2 from $(\omega_1,\in)$ into $(\mathbb{N}^{\mathbb{N}},<_F)$ with the properties specified above such that $s = \langle \text{ran } \pi_1, \text{ran } \pi_2 \rangle$. Set $t = (\pi_1,\pi_2)$, and let $\mathbb{Q}(t)$ be as described. The Boolean algebra $\mathbb{B}$ is the completion of $\mathbb{Q}(t)$. As we noted after 6.11, $|Q(t)| = \aleph_1$ and $\mathbb{Q}(t)$ is separative, and so $|B| \geq \aleph_1$. By 5.16, $|B| \leq 2^{\aleph_0}$, and so, with CH, $|B| = \aleph_1$. The result follows from 8.17. ∎

We now prove a lemma by using the iteration theorem. This lemma is a basic tool to be used the construction of a complete Boolean algebra $\mathbb{A}$ with $[\![\text{MA}]\!]^{\mathbb{A}} = 1$, for it shows that we can eliminate a counter-example to MA by passing to a larger Boolean algebra.

8.19 LEMMA (CH)

Let $\mathbb{B}$ be a complete ccc Boolean algebra of cardinality $\aleph_1$, let $p \in V^{\mathbb{B}}$ be such that

$$[\![\text{'}p \text{ is a ccc partially ordered set of cardinality } \aleph_1\text{'}]\!]^{\mathbb{B}} = 1,$$

and let $d \in V^{\mathbb{B}}$ be such that

$$[\![\text{'}d \text{ is a family of dense subsets of } p\text{'}]\!]^{\mathbb{B}} = 1.$$

Then there is a complete ccc Boolean algebra $\mathbb{C}$ of cardinality $\aleph_1$ containing $\mathbb{B}$ as a complete subalgebra such that

$$[\![\text{'there is a } d\text{-generic filter in } p\text{'}]\!]^{\mathcal{C}} = 1. \quad (16)$$

Proof

By the fullness lemma, there is a term u in $V^{\mathcal{B}}$ such that

$$[\![\text{'}u \text{ is the completion of } p\text{'}]\!]^{\mathcal{B}} = 1.$$

By 8.16,

$$[\![\text{'}[\![\text{'there is a } (d)^{\vee}_{u}\text{-generic filter in } (p)^{\vee}_{u}\text{'}]\!]^{u} = 1\text{'}]\!]^{\mathcal{B}} = 1,$$

where we are using the notation of 8.14. By 8.14(i), there is a complete Boolean algebra $\mathcal{C}$ containing $\mathcal{B}$ as a complete subalgebra such that (16) holds. By 5.16,

$$[\![\text{'}u \text{ is ccc and has cardinality } \leqslant 2^{\aleph_0}\text{'}]\!]^{\mathcal{B}} = 1,$$

and so, by 8.14(ii), we can suppose that $\mathcal{C}$ is ccc and has cardinality $2^{\aleph_0}$. With CH, $\mathcal{C}$ has cardinality $\aleph_1$. ∎

Let $\langle A_1,B_1\rangle$ and $\langle A_2,B_2\rangle$ be pregaps in $(P,<)$. Then $\langle A_1,B_1\rangle$ and $\langle A_2,B_2\rangle$ are <u>incompatible</u> if either $b \leqslant a$ for some $a \in A_1$ and $b \in B_2$ or $b \leqslant a$ for some $a \in A_2$ and $b \in B_1$. In this case there is no element of P which interpolates both $\langle A_1,B_1\rangle$ and $\langle A_2,B_2\rangle$.

8.20 LEMMA (CH)

Let $\mathcal{B}$ be a complete ccc Boolean algebra of cardinality $\aleph_1$, and let $t \in V^{\mathcal{B}}$ be such that

$$[\![\text{'}t \text{ is an embedding of } (\mathbb{Q},<)^{\vee} \text{ into } (\mathbb{N}^{\mathbb{N}},<_F)\text{'}]\!]^{\mathcal{B}} = 1.$$

Then there is a complete ccc Boolean algebra $\mathcal{C}$ of cardinality $\aleph_1$ containing $\mathcal{B}$ as a complete subalgebra such that

$$[\![\text{'the range of } t \text{ contains a Hausdorff gap'}]\!]^{\mathcal{C}} = 1.$$

<u>Proof</u>

The set J was defined on page 121. Let $x \in J$, and let $\langle A_x, B_x \rangle$ be the corresponding pregap, defined on page 121. By 6.22(iv), $\langle A_x, B_x \rangle$ is an $(\aleph_1, \aleph_1)$-gap in $(\mathbb{Q}, <)$, and so, since $\mathcal{B}$ is ccc,

$$[\![\text{'} \langle A_x, B_x \rangle^\vee \text{ is an } (\aleph_1, \aleph_1)\text{-pregap in } (\mathbb{Q}, <)^\vee \text{'}]\!]^{\mathcal{B}} = 1.$$

By the fullness lemma, there is a term s_x in $V^{\mathcal{B}}$ such that

$$[\![\text{'} s_x \text{ is the } (\aleph_1, \aleph_1)\text{-pregap in } (\mathbb{N}^{\mathbb{N}}, <_{\mathcal{F}}) \text{ given by the image } \langle A_x, B_x \rangle^\vee \text{ under } t \text{'}]\!]^{\mathcal{B}} = 1.$$

If $x, y \in J$ and $x \neq y$, then $\langle A_x, B_x \rangle$ and $\langle A_y, B_y \rangle$ are incompatible, and so

$$[\![\text{'} s_x \text{ and } s_y \text{ are incompatible in } (\mathbb{N}^{\mathbb{N}}, <_{\mathcal{F}}) \text{'}]\!]^{\mathcal{B}} = 1. \quad (17)$$

Set

$$H = \{x \in J : [\![\text{'} s_x \text{ is a gap in } (\mathbb{N}^{\mathbb{N}}, <_{\mathcal{F}}) \text{'}]\!]^{\mathcal{B}} = 1\}.$$

We *claim* that $|J \setminus H| \leqslant \aleph_1$. For suppose not. Since $\mathcal{B}$ has cardinality $\aleph_1$, there exists $b \in B \setminus \{0\}$ and $C \subset J \setminus H$ with $|C| = \aleph_2$ such that, for each $x \in C$,

$$[\![\text{'} s_x \text{ can be interpolated in } (\mathbb{N}^{\mathbb{N}}, <_{\mathcal{F}}) \text{'}]\!]^{\mathcal{B}} = b.$$

By the fullness lemma, for each $x \in C$ there exists $f_x \in V^{\mathcal{B}}$ such that

$$[\![\text{'} f_x \in \mathbb{N}^{\mathbb{N}} \text{ and } f_x \text{ interpolates } s_x \text{'}]\!]^{\mathcal{B}} = b.$$

By 7.29(ii), $[\![\mathrm{CH}]\!]^{\mathcal{B}} = 1$, and $[\![\text{'} \check{\aleph}_1 = \aleph_1 \text{'}]\!]^{\mathcal{B}} = 1$ because $\mathcal{B}$ is ccc, and so $[\![\text{'} |\mathbb{N}^{\mathbb{N}}| = \check{\aleph}_1 \text{'}]\!]^{\mathcal{B}} = 1$. Thus, by many applications of the fullness lemma, there is a sequence $\langle f_\alpha : \alpha < \omega_1 \rangle$ of terms in $V^{\mathcal{B}}$ such that, for each $f \in V^{\mathcal{B}}$,

$$[\![\text{'}f \in \mathbb{N}^{\mathbb{N}}\text{'}]\!]^{\mathcal{B}} = \bigvee\{[\![f \mathrel{\hat{=}} f_\alpha]\!]^{\mathcal{B}} : \alpha < \omega_1\}.$$

So, for each $x \in C$, $\bigvee\{[\![f_x \mathrel{\hat{=}} f_\alpha]\!]^{\mathcal{B}} : \alpha < \omega_1\} \geqslant b$. But $\mathcal{B}$ has cardinality $\aleph_1$ and $|C| = \aleph_2$, and so there exist $x, y \in C$ with $x \neq y$ and $\alpha < \omega_1$ such that

$$[\![f_x \mathrel{\hat{=}} f_\alpha]\!]^{\mathcal{B}} = [\![f_y \mathrel{\hat{=}} f_\alpha]\!]^{\mathcal{B}} \text{ and } [\![f_x \mathrel{\hat{=}} f_\alpha]\!]^{\mathcal{B}} \wedge b \neq 0.$$

Thus $[\![f_x \mathrel{\hat{=}} f_y]\!]^{\mathcal{B}} \wedge b \neq 0$, and so

$$[\![\text{'}f_x \text{ interpolates } s_x\text{'}]\!]^{\mathcal{B}} \wedge [\![\text{'}f_y \text{ interpolates } s_y\text{'}]\!]^{\mathcal{B}} \wedge [\![f_x \mathrel{\hat{=}} f_y]\!]^{\mathcal{B}} \neq 0,$$

a contradiction of (17). Thus the claim holds.

By 6.22(i), $|J| = 2^{\aleph_1}$, and so $H \neq \emptyset$.

Take $x \in H$. By 8.18 and the fullness lemma, there is a term u in $V^{\mathcal{B}}$ with

$$[\![\text{'}u \text{ is a complete ccc Boolean algebra of cardinality } \aleph_1\text{'}]\!]^{\mathcal{B}} = 1$$

and

$$[\![\text{'}[\![\text{'}(s_x)^{\vee}_{u} \text{ is a Hausdorff gap'}]\!]^{u} = 1\text{'}]\!]^{\mathcal{B}} = 1.$$

By the iteration theorem 8.14(i), there is a complete Boolean algebra $\mathcal{C}$ containing $\mathcal{B}$ as a complete subalgebra such that

$$[\![\text{'}s_x \text{ is a Hausdorff gap'}]\!]^{\mathcal{C}} = 1;$$

by 8.14(ii), we can suppose that $\mathcal{C}$ is ccc and has cardinality $\aleph_1$.

Since the relevant properties can be expressed by Δ_0-formulae, it follows that

$$[\![\text{'}t \text{ is an embedding of } (\mathbb{Q},<)^{\vee} \text{ into } (\mathbb{N}^{\mathbb{N}},<_F)\text{'}]\!]^{\mathcal{C}} = 1$$

and that

$$[\![\text{'}s_x \text{ is a pregap in the range of } t\text{'}]\!]^{\mathbb{C}} = 1.$$

Thus $\mathbb{C}$ has the required properties. ∎

Suppose that $f : \text{dom } f \to \{0,1\}$ is a function such that: (i) $\text{dom } f$ is an ordinal; (ii) f is eventually constant; (iii) $\{\alpha : f(\alpha) = 0\}$ is a non-empty, proper subset of $\text{dom } f$ with no maximum element. Now each of these properties of f can be specified by Δ_o-formulae, and the set $\mathbb{Q}$ is just the set of functions f satisfying these properties with $\text{dom } f = \omega_1$ (cf. Definition 6.21). Let $\mathbb{B}$ be a complete Boolean algebra, and let $x \in \mathbb{Q}$. Then

$$[\![\text{'}\check{x} \in \mathbb{Q}\text{'}]\!]^{\mathbb{B}} = [\![\text{'dom } \check{x} = \omega_1\text{'}]\!]^{\mathbb{B}} = [\![\text{'}\check{\omega}_1 = \omega_1\text{'}]\!]^{\mathbb{B}}.$$

Since the order on $\mathbb{Q}$ can be specified by a Δ_o-formula, it follows that, for each complete ccc Boolean algebra $\mathbb{B}$,

$$[\![\text{'}(\mathbb{Q},<)^{\vee} \text{ is a suborder of } (\mathbb{Q},<)\text{'}]\!]^{\mathbb{B}} = 1. \qquad (18)$$

Recall that it follows from the main result of Chapter 6, Theorem 6.25, that, to build a model of ZFC + NDH, it suffices to build a model of ZFC + MA + ¬CH in which there is no embedding of $(\mathbb{R},<)$ into $(\mathbb{N}^{\mathbb{N}},<_F)$. By 6.22(ii), each $(\aleph_1,\aleph_1)$-pregap in $(\mathbb{Q},<)$ is interpolated in $(\mathbb{R},<)$. We thus see that, if $\mathbb{B}$ is a complete ccc Boolean algebra, and if

$$[\![\text{'the range of each embedding of } (\mathbb{Q},<)^{\vee} \text{ into } (\mathbb{N}^{\mathbb{N}},<_F) \text{ contains a Hausdorff gap'}]\!]^{\mathbb{B}} = 1, \qquad (19)$$

then

$$[\![\text{'}(\mathbb{R},<) \text{ does not embed in } (\mathbb{N}^{\mathbb{N}},<_F)\text{'}]\!]^{\mathbb{B}} = 1. \qquad (20)$$

To build $\mathbb{B}$, we shall construct a sequence $\langle \mathbb{B}_\alpha : \alpha < \omega_2 \rangle$ of complete ccc Boolean algebras such that, if $\alpha < \beta < \omega_2$, then $\mathbb{B}_\alpha$ is a complete subalgebra of $\mathbb{B}_\beta$. If γ is a limit ordinal with $\gamma < \omega_2$, we shall take $\mathbb{B}_\gamma$ to be the completion of $\bigcup_{\alpha<\gamma} \mathbb{B}_\alpha$, and we shall set $\mathbb{B} = \bigcup_{\alpha<\omega_2} \mathbb{B}_\alpha$. By 8.15, $\mathbb{B}$ will be a complete ccc Boolean algebra. We shall assume CH, and we shall choose each $\mathbb{B}_\alpha$ to be of cardinality $\aleph_1$.

Suppose that $\mathbb{B}_\alpha$ has been constructed and that r is a term in $V^{\mathbb{B}_\alpha}$ such that

$$[\![\text{'}r \text{ is an embedding of } (\mathbb{Q},<)^\vee \text{ into } (\mathbb{N}^{\mathbb{N}},<_F)\text{'}]\!]^{\mathbb{B}_\alpha} = 1.$$

Then by 8.20, there is a complete ccc Boolean algebra $\mathbb{B}_{\alpha+1}$ of cardinality $\aleph_1$ containing $\mathbb{B}_\alpha$ as a complete subalgebra such that

$$[\![\text{'the range of } r \text{ contains a Hausdorff gap'}]\!]^{\mathbb{B}_{\alpha+1}} = 1.$$

The key point is that, if $\mathbb{C}$ is *any* complete ccc Boolean algebra containing $\mathbb{B}_{\alpha+1}$ as a complete subalgebra, then it follows from (13) that

$$[\![\text{'the range of } r \text{ contains a Hausdorff gap'}]\!]^{\mathbb{C}} = 1.$$

Thus, if r has been eliminated at stage $\alpha + 1$, it can never trouble us again.

In a similar fashion, we can use 8.19 to eliminate in $V^{\mathbb{B}_{\alpha+1}}$ a potential counter-example to MA which occurs in some $V^{\mathbb{B}_\alpha}$; we shall again see that, if $\mathbb{C}$ is any complete ccc Boolean algebra containing $\mathbb{B}_{\alpha+1}$ as a complete subalgebra, then the counter-example remains eliminated in $V^{\mathbb{C}}$.

One problem still remains: even though $\mathbb{B} = \bigcup_{\alpha<\omega_2} \mathbb{B}_\alpha$, it does not follow that $V^{\mathbb{B}} = \bigcup_{\alpha<\omega_2} V^{\mathbb{B}_\alpha}$. But we shall show that all the terms in $V^{\mathbb{B}}$ that concern us are equivalent to terms in $\bigcup_{\alpha<\omega_2} V^{\mathbb{B}_\alpha}$. We must deal with: (i) terms for partially ordered sets with underlying set ω_1; (ii) terms for families having cardinality $\aleph_1$ of subsets of ω_1; (iii) terms for embeddings from $(\mathbb{Q},<)^\vee$ into $(\mathbb{N}^{\mathbb{N}},<_F)$. In dealing with these cases, we shall fix some notation that will be used in the proof of the main theorem.

Let $\mathbf{B}$ be a complete ccc Boolean algebra.

Let $F : \omega_1 \times \omega_1 \to B$ be a function. Define a term e_F in $V^{\mathbf{B}}$ by

$$\mathrm{dom}\, e_F = \{(\alpha,\beta)^\vee : \alpha,\beta < \omega_1\},\ e_F((\alpha,\beta)^\vee) = F(\alpha,\beta).$$

Let $H : \omega_1 \times \omega_1 \to B$ be a function. For each $\zeta < \omega_1$, define a term d_ζ^H in $V^{\mathbf{B}}$ by

$$\mathrm{dom}\, d_\zeta^H = \{\check{\eta} : \eta < \omega\},\ \ d_\zeta^H(\check{\eta}) = H(\zeta,\eta),$$

and define a term D_H in $V^{\mathbf{B}}$ by

$$\mathrm{dom}\, D_H = \{d_\zeta^H : \zeta < \omega_1\},\ D_H(d_\zeta^H) = 1$$

(so that $D_H = \{d_\zeta^H : \zeta < \omega_1\}^{\#}$ in the notation of 7.40).

Let $J : \mathbb{Q} \times \mathbb{N} \times \mathbb{N} \to B$ be a function. For each $x \in \mathbb{Q}$, define a term f_x^J in $V^{\mathbf{B}}$ by

$$\mathrm{dom}\, f_x^J = \{(m,n)^\vee : m,n \in \mathbb{N}\},\ \ f_x^J((m,n)^\vee) = J(x,m,n),$$

and define a term t_J in $V^{\mathbf{B}}$ by

$$t_J = \{\{\{\check{x}\}^{\#},\{\check{x},f_x\}^{\#}\}^{\#} : x \in \mathbb{Q}\}^{\#}.$$

8.21 LEMMA

Let $\mathbf{B}$ be a complete Boolean algebra.

(i) For each $p \in V^{\mathbf{B}}$ such that

$$[\![\,'p \text{ is a partially ordered set with underlying set } \check{\aleph}_1{}'\,]\!]^{\mathbf{B}} = 1,$$

there is a function $F : \omega_1 \times \omega_1 \to B$ such that

$$[\![\text{'}p \text{ is equal to } (\check{\omega}_1, e_F)\text{'}]\!]^{\mathbb{B}} = 1.$$

(ii) For each $D \in V^{\mathbb{B}}$ such that

$$[\![\text{'}D \subset P(\check{\omega}_1) \text{ and } |D| \leqslant \check{\aleph}_1\text{'}]\!]^{\mathbb{B}} = 1,$$

there is a function $H : \omega_1 \times \omega_1 \to B$ such that

$$[\![D \,\hat{=}\, D_H]\!]^{\mathbb{B}} = 1.$$

(iii) For each $t \in V^{\mathbb{B}}$ such that

$$[\![\text{'}t \text{ is an embedding from } (\mathbb{Q},<)^{\vee} \text{ into } (\mathbb{N}^{\mathbb{N}},<_F)\text{'}]\!]^{\mathbb{B}} = 1,$$

there is a function $J : \mathbb{Q} \times \mathbb{N} \times \mathbb{N} \to B$ such that

$$[\![t \,\hat{=}\, t_J]\!]^{\mathbb{B}} = 1.$$

<u>Proof</u>

(i) Since $\mathbb{B}$ is ccc, $[\![\text{'}\check{\aleph}_1 = \aleph_1\text{'}]\!]^{\mathbb{B}} = 1$, and so, by the fullness lemma, there is a term $e \in V^{\mathbb{B}}$ with

$$[\![\text{'}p \text{ is equal to the partially ordered set } (\check{\omega}_1, e)\text{'}]\!]^{\mathbb{B}} = 1.$$

Define $F : (\alpha,\beta) \mapsto [\![(\alpha,\beta)^{\vee} \,\hat{\in}\, e]\!]^{\mathbb{B}}$, $\omega_1 \times \omega_1 \to B$. Then $[\![e \,\hat{=}\, e_F]\!]^{\mathbb{B}} = 1$, and the result follows.

(ii) By the fullness lemma, there is a term $t \in V^{\mathbb{B}}$ with

$$[\![\text{'}t \text{ is a function from } \check{\omega}_1 \text{ onto } D\text{'}]\!]^{\mathbb{B}} = 1,$$

and, for each $\zeta < \omega_1$, there are terms $s_\zeta \in V^{\mathbb{B}}$ with

$$[\![\text{'}s_\zeta \text{ is the value of } t \text{ at } \check{\zeta}\text{'}]\!]^{\mathbb{B}} = 1.$$

Define $H : (\zeta,\eta) \mapsto [\![\check{\eta} \hat{\in} s_\zeta]\!]^{\mathbb{B}}$, $\omega_1 \times \omega_1 \to B$. Then, for each $\zeta < \omega_1$, $[\![s_\zeta \hat{=} d^H_\zeta]\!]^{\mathbb{B}} = 1$, and so $[\![D \hat{=} D_H]\!]^{\mathbb{B}} = 1$.

(iii) Define $J : \mathbb{Q} \times \mathbb{N} \times \mathbb{N} \to B$ by

$$J(x,m,n) = [\![\text{'the value of } t \text{ at } \check{x} \text{ takes the value } \check{n} \text{ at } \check{m}\text{'}]\!]^{\mathbb{B}}.$$

From the above definitions of t_J and of f^J_x,

$$[\![\text{'}t_J \text{ is a function with domain } \check{\mathbb{Q}}\text{'}]\!]^{\mathbb{B}} = 1$$

and

$$[\![\text{'the value of } t_J \text{ at } \check{x} \text{ is } f^J_x\text{'}]\!]^{\mathbb{B}} = 1.$$

It is now routine to check that $[\![t \hat{=} t_J]\!]^{\mathbb{B}} = 1$. ∎

Note that, if $\mathbb{B}$ in the above lemma is ccc, then $[\![\text{'}\omega_1 = \check{\omega}_1\text{'}]\!]^{\mathbb{B}} = 1$, and so $\check{\omega}_1$ and $\check{\aleph}_1$ can be replaced by ω_1 and $\aleph_1$, respectively, in expressions $[\![\text{'...'}]\!]^{\mathbb{B}}$.

We now come to our main theorem.

8.22 THEOREM ($2^{\aleph_0} = \aleph_1$ and $2^{\aleph_1} = \aleph_2$)
There is a complete Boolean algebra $\mathbb{A}$ such that

$$[\![\mathrm{MA} + \mathrm{NDH}]\!]^{\mathbb{A}} = 1.$$

Proof

For $\omega_1 \leqslant \zeta \leqslant \omega_2$, set $S_\zeta = \zeta^{\omega_1} \times \zeta^{\omega_1}$ and set $T_\zeta = \zeta^{\mathbb{Q} \times \mathbb{N} \times \mathbb{N}}$ (where we recall that x^y denotes the set of functions from y to x). Write $S = S_{\omega_2}$ and $T = T_{\omega_2}$. Then

$$S = \bigcup\{S_\zeta : \zeta < \omega_2\}, \quad T = \bigcup\{T_\zeta : \zeta < \omega_2\}.$$

Also, for $\zeta < \omega_2$, $|S_\zeta| = \aleph_1^{\aleph_1} = 2^{\aleph_1}$, and so, by hypothesis, $|S_\zeta| = \aleph_2$. Similarly, $|T_\zeta| = \aleph_1^{\aleph_1} = \aleph_2$ because

$|\mathbb{Q}| = 2^{\aleph_0} = \aleph_1$. Thus $|S| = |T| = \aleph_2$. Fix an enumeration $\langle z_\alpha : \alpha < \omega_2 \rangle$ of $S \cup T$ such that $\{\alpha : z_\alpha = x\}$ is cofinal in ω_2 for each $x \in S \cup T$: this is possible because $|\omega_2| = |\omega_2 \times \omega_2|$.

We shall define $\langle \mathbb{A}_\alpha : \alpha < \omega_2 \rangle$ to be a sequence of complete ccc Boolean algebras of cardinality $\aleph_1$ such that, if $\alpha < \beta < \omega_2$, then $\mathbb{A}_\alpha$ is a complete subalgebra of $\mathbb{A}_\beta$, and such that, if β is a limit ordinal with $\beta < \omega_2$, then $\mathbb{A}_\alpha$ is the completion of $\bigcup_{\alpha<\beta} \mathbb{A}_\alpha$. Further, each underlying set A_α is chosen to be a subset of ω_2. The definition is by recursion on α, and it is guided by the enumeration $\langle z_\alpha : \alpha < \omega_2 \rangle$, which is used to "anticipate counter-examples".

Choose $\mathbb{A}_o$ to be a complete ccc Boolean algebra of cardinality $\aleph_1$ with $A_o \subset \omega_2$.

Suppose now that $\beta < \omega_2$ and that $\langle \mathbb{A}_\alpha : \alpha \leqslant \beta \rangle$ has been defined. We define $\mathbb{A}_{\beta+1}$. There are two cases.

Case 1. Suppose that $z_\beta \in S$, say $z_\beta = (F,H)$, where $F,H \in \omega_2^{\omega_1}$.

If $\operatorname{ran} F \cup \operatorname{ran} H \not\subset A_\beta$, set $\mathbb{A}_{\beta+1} = \mathbb{A}_\beta$.

If $\operatorname{ran} F \cup \operatorname{ran} H \subset A_\beta$, let e_F and D_H be the corresponding terms to F and H, as above. Set

$$a = [\![\, '(\check{\omega}_1, e_F) \text{ is a ccc partially ordered set}' \,]\!]^{\mathbb{A}_\beta},$$
$$b = [\![\, 'D_H \text{ is a family of dense subsets of } (\check{\omega}_1, e_F)' \,]\!]^{\mathbb{A}_\beta}.$$

If $a < 1$ or $b < 1$, then again set $\mathbb{A}_{\beta+1} = \mathbb{A}_\beta$. Finally, suppose that $a = b = 1$. By 8.19, there is a complete ccc Boolean algebra $\mathbb{C}$ of cardinality $\aleph_1$ containing $\mathbb{A}_\beta$ as a complete subalgebra such that

$$[\![\, '\text{there is a } D_H\text{-generic filter in } (\check{\omega}_1, e_F)' \,]\!]^{\mathbb{C}} = 1:$$

set $\mathbb{A}_{\beta+1} = \mathbb{C}$. Since $|A_\beta| = \aleph_1$, $|\omega_2 \setminus A_\beta| = \aleph_2$, and so we can suppose that $A_{\beta+1} \subset \omega_2$.

<u>Case 2</u>. Suppose that $z_\beta \in T$, say $z_\beta = J$, where $J \in \omega_2^{\mathbb{Q} \times \mathbb{N} \times \mathbb{N}}$.

If $\operatorname{ran} J \not\subset A_\beta$, set $\mathbb{A}_{\beta+1} = \mathbb{A}_\beta$.

If $\operatorname{ran} J \subset A_\beta$, let t_J be the corresponding term to J, as above. Set

$$a = [\![\text{'}t_J \text{ is an embedding of } (\mathbb{Q},<)^\vee \text{ into } (\mathbb{N}^\mathbb{N},<_F)\text{'}]\!]^{\mathbb{A}_\beta}.$$

If $a < 1$, then again set $\mathbb{A}_{\beta+1} = \mathbb{A}_\beta$. Finally, suppose that $a = 1$. By 8.20, there is a complete ccc Boolean algebra $\mathbb{C}$ of cardinality $\aleph_1$ containing $\mathbb{A}_\beta$ as a complete subalgebra such that

$$[\![\text{'the range of } t_J \text{ contains a Hausdorff gap'}]\!]^{\mathbb{C}} = 1:$$

set $\mathbb{A}_{\beta+1} = \mathbb{C}$. Again, we can suppose that $A_{\beta+1} \subset \omega_2$.

Suppose that $\beta < \omega_2$, that β is a limit ordinal, and that $\mathbb{A}_\alpha$ has been defined for $\alpha < \beta$. Set $\mathbb{C} = \bigcup_{\alpha<\beta} \mathbb{A}_\alpha$. By 8.15, $\mathbb{C}$ is a ccc Boolean algebra, and clearly $\mathbb{C}$ has cardinality $\aleph_1$. Let $\mathbb{A}_\beta$ be a complete Boolean algebra containing $\mathbb{C}$ as a dense subalgebra. Then $\mathbb{A}_\beta$ is isomorphic to the completion of $\mathbb{C}$, and so, by 5.16, $\mathbb{A}_\beta$ has cardinality $\aleph_1$. We may suppose that $A_\beta \subset \omega_2$.

This completes the definition of the sequence $\langle \mathbb{A}_\alpha : \alpha < \omega_2 \rangle$.

Set $\mathbb{A} = \bigcup_{\alpha<\omega_2} \mathbb{A}_\alpha$. By 8.15, $\mathbb{A}$ is a complete ccc Boolean algebra, and clearly $|A| \leqslant \aleph_2$.

The next step is to prove that $[\![MA + \neg CH]\!]^{\mathbb{A}} = 1$.

We first detour to note that, if $2^{\aleph_0} \leqslant \aleph_2$, then $MA + \neg CH$ is equivalent to the following.

Let $<$ be a partial order on ω_1 such that $(\omega_1,<)$ is ccc. Suppose that $\mathcal{D}$ is a family of dense subsets of $(\omega_1,<)$ with $|\mathcal{D}| \leqslant \aleph_1$. Then there is a $\mathcal{D}$-generic filter in $(\omega_1,<)$. (*)

To see this, first observe that, trivially, $MA + \neg CH$ implies (*). Secondly, observe that (*) is equivalent to the assertion that, if $\mathbb{P}$ is a ccc partial order with $|P| \leq \aleph_1$, and if $\mathcal{D}$ is a family of dense subsets P with $|\mathcal{D}| \leq \aleph_1$, then there is a $\mathcal{D}$-generic filter in $\mathbb{P}$.

Thirdly, we prove that this assertion implies $\neg CH$. For take $X \subset \mathbb{R}$ with $|X| = \aleph_1$, and let P be the collection of rational intervals of $\mathbb{R}$. Then $(P,\subset)$ is a ccc partial ordered set with $|P| = \aleph_0 < \aleph_1$. For $x \in X$, set $d_x = \{I \in P : x \notin \bar{I}\}$, and set $\mathcal{D} = \{d_x : x \in X\}$. Then $\mathcal{D}$ is a family of dense subsets of $(P,\subset)$ with $|\mathcal{D}| = \aleph_1$. By the assertion, there is a $\mathcal{D}$-generic filter, say F, in $(P,\subset)$, and $\cap\{\bar{I} : I \in F\}$ is a non-empty subset of $\mathbb{R}$ disjoint from X. Thus $X \neq \mathbb{R}$, and CH fails.

Thus, given that $2^{\aleph_0} \leq \aleph_2$, (*) implies that $2^{\aleph_0} = \aleph_2$, and so, by 5.18, the assertion implies MA', and hence, by 5.20, MA.

We now return to our main proof. To obtain a contradiction, suppose that $[\![MA + \neg CH]\!]^{\mathbb{A}} \neq 1$. We know that $|A| \leq \aleph_2$, and so, by 7.29(i), $[\!['2^{\aleph_0} \leq \aleph_2']\!]^{\mathbb{A}} = 1$. Further, since $\mathbb{A}$ is ccc, $[\!['\check{\omega}_1 = \omega_1']\!]^{\mathbb{A}} = 1$. Hence, by the above formulation (*) of $MA + \neg CH$, there are terms e and D in $V^{\mathbb{A}}$ and $a \in A\backslash\{0\}$ with:

(i) $[\!['(\check{\omega}_1, e)$ is a ccc partially ordered set$']\!]^{\mathbb{A}} \geq a$;

(ii) $[\!['$each element of D is dense in $(\check{\omega}_1, e)']\!]^{\mathbb{A}} \geq a$;

(iii) $[\!['D$ has cardinality $\leq \check{\aleph}_1']\!]^{\mathbb{A}} \geq a$;

(iv) $[\!['$there is no D-generic filter on $(\check{\omega}_1, e)']\!]^{\mathbb{A}} \geq a$.

By the mixing lemma, we can suppose that the Boolean values specified in the left-hand sides of (i) - (iii) are actually 1.

By Lemma 8.21, there are functions $F : \omega_1 \times \omega_1 \to A$ and $H : \omega_1 \times \omega_1 \to A$ such that

$$[\![\text{'}(\check{\omega}_1, e) \text{ is equal to } (\check{\omega}_1, e_F)\text{'}]\!]^{\mathbb{A}} = [\![D \mathrel{\hat{=}} D_H]\!]^{\mathbb{A}} = 1.$$

We have

$$[\![\text{'}(\check{\omega}_1, e_F) \text{ is a ccc partially ordered set'}]\!]^{\mathbb{A}} = 1,$$

$$[\![\text{'each element of } D_H \text{ is dense in } (\check{\omega}_1, e_F)\text{'}]\!]^{\mathbb{A}} = 1,$$

$$[\![\text{'}D_H \text{ has cardinality} \leq \check{\aleph}_1\text{'}]\!]^{\mathbb{A}} = 1.$$

Since $A \subset \omega_2$ and $\{\alpha : z_\alpha = (F,H)\}$ is cofinal in ω_2, there exists $\beta < \omega_2$ with $\operatorname{ran} F \cup \operatorname{ran} H \subset A_\beta$ and $z_\beta = (F,H)$. Clearly, $e_F, D_H \in V^{\mathbb{A}_\beta}$, and $\mathbb{A}_\beta$ is a complete subalgebra of $\mathbb{A}$. Thus, by (14),

$$[\![\text{'}(\check{\omega}_1, e_F) \text{ is a ccc partially ordered set'}]\!]^{\mathbb{A}_\beta} = 1;$$

since the relevant statement about D_H, $\check{\omega}_1$, and e_F is specified by a Δ_0-formula,

$$[\![\text{'each element of } D_H \text{ is dense in } (\check{\omega}_1, e_F)\text{'}]\!]^{\mathbb{A}_\beta} = 1;$$

since cardinals are preserved from $V^{\mathbb{A}_\beta}$ to $V^{\mathbb{A}}$,

$$[\![\text{'}D_H \text{ has cardinality } \leq \check{\aleph}_1\text{'}]\!]^{\mathbb{A}_\beta} = 1.$$

By case 1 of the construction and by the fullness lemma, there is a term $t \in V^{\mathbb{A}_{\beta+1}}$ such that

$$[\![\text{'}t \text{ is a } D_H\text{-generic filter in } (\check{\omega}_1, e_F)\text{'}]\!]^{\mathbb{A}_{\beta+1}} = 1.$$

Again, since the relevant statement is specified by a Δ_0-formula and since $\mathbb{A}_{\beta+1}$ is a complete subalgebra of $\mathbb{A}$,

$$[\![\text{'}t \text{ is a } D_H\text{-generic filter in } (\check{\omega}_1, e_F)\text{'}]\!]^{\mathbb{A}} = 1,$$

and so

$$[\![\text{'}t \text{ is a } D\text{-generic filter in } (\check{\omega}_1, e)\text{'}]\!]^{\mathbb{A}} = 1.$$

But this is a contradiction of (iv). Thus it is true that $[\![\mathrm{MA} + \neg\mathrm{CH}]\!]^{\mathbb{A}} = 1$.

Our final step is to prove that $[\![\mathrm{NDH}]\!]^{\mathbb{A}} = 1$. Since $[\![\mathrm{MA} + \neg\mathrm{CH}]\!]^{\mathbb{A}} = 1$, it is sufficient to prove that

$$[\![\text{'the range of each embedding of } (\mathbb{Q},<)^{\vee} \text{ into } (\mathbb{N}^{\mathbb{N}},<_F) \text{ contains a Hausdorff gap'}]\!]^{\mathbb{A}} = 1 \tag{21}$$

For suppose that (21) holds. Since (19) implies (20),

$$[\![\text{'}(\mathbb{R},<) \text{ does not embed in } (\mathbb{N}^{\mathbb{N}},<_F)\text{'}]\!]^{\mathbb{A}} = 1,$$

and so, by Theorem 6.25, $[\![\mathrm{NDH}]\!]^{\mathbb{A}} = 1$.

To obtain a contradiction, suppose that (21) fails. Then there is a term t in $V^{\mathbb{A}}$ and $a \in A\backslash\{0\}$ with

$$[\![\text{'}t \text{ is an embedding of } (\mathbb{Q},<)^{\vee} \text{ into } (\mathbb{N}^{\mathbb{N}},<_F) \text{ whose range does not contain a Hausdorff gap'}]\!]^{\mathbb{A}} \geq a. \tag{22}$$

Now we proved in Chapter 6 that $(\mathbb{Q},<)$ does embed in $(\mathbb{N}^{\mathbb{N}},<_F)$, and so, using (18), we have

$$[\![\text{'there is an embedding of } (\mathbb{Q},<)^{\vee} \text{ into } (\mathbb{N}^{\mathbb{N}},<_F)\text{'}]\!]^{\mathbb{A}} = 1.$$

It follows from the mixing and fullness lemmas that we can suppose that

$$[\![\text{'}t \text{ is an embedding of } (\mathbb{Q},<)^{\vee} \text{ into } (\mathbb{N}^{\mathbb{N}},<_F)\text{'}]\!]^{\mathbb{A}} = 1.$$

By Lemma 8.21(iii), there is a function $J : \mathbb{Q} \times \mathbb{N} \times \mathbb{N} \to A$ such that $[\![t \,\hat{=}\, t_J]\!]^{\mathbb{A}} = 1$. We have

$$[\![\text{'}t_J \text{ is an embedding of } (\mathbb{Q},<)^{\vee} \text{ into } (\mathbb{N}^{\mathbb{N}},<_F)\text{'}]\!]^{\mathbb{A}} = 1.$$

As before, there exists $\beta < \omega_2$ with $\operatorname{ran} J \subset A_\beta$ and $z_\beta = J$. Clearly $t_J \in V^{\mathbb{A}_\beta}$, and $\mathbb{A}_\beta$ is a complete subalgebra of $\mathbb{A}$, and so

$$[\![\text{'} t_J \text{ is an embedding of } (\mathbb{Q},<)^\vee \text{ into } (\mathbb{N}^{\mathbb{N}},<_F)]\!]^{\mathbb{A}_\beta} = 1.$$

By case 2 of the construction and by the fullness lemma, there is a term $s \in V^{\mathbb{A}_{\beta+1}}$ such that

$$[\![\text{'} s \text{ is a Hausdorff gap in the range of } t_J \text{'}]\!]^{\mathbb{A}_{\beta+1}} = 1.$$

Since "s is a pregap in $(\mathbb{N}^{\mathbb{N}},<_F)$ in the range of t_J" is specified by a Δ_o-formula, it follows from (13) that

$$[\![\text{'} s \text{ is a Hausdorff gap in the range of } t_J \text{'}]\!]^{\mathbb{A}} = 1.$$

But this is a contradiction of (22). Thus the final claim is established.

This completes the proof of the theorem. ∎

The next theorem is obtained by combining 7.37 with 8.14, but the use of the iteration theorem here is rather "heavy-handed".

8.23 THEOREM

There is a complete Boolean algebra $\mathbb{C}$ such that

$$[\![\text{'} 2^{\aleph_o} = \aleph_1 \text{ and } 2^{\aleph_1} = \aleph_2 \text{'}]\!]^{\mathbb{C}} = 1.$$

<u>Proof</u>

By 7.37, there is a complete Boolean algebra $\mathbb{B}$ such that $[\![CH]\!]^{\mathbb{B}} = 1$. By 7.39, and the fullness lemma, there is a term u in $V^{\mathbb{B}}$ such that

$[\![$'u is a complete Boolean algebra'$]\!]^{\mathbb{B}}$

$$= [\![\text{'}[\![2^{\aleph_o} = \aleph_1 \text{ and } 2^{\aleph_1} = \aleph_2]\!]^u = 1\text{'}]\!]^{\mathbb{B}} = 1.$$

The result follows from 8.14(i). ∎

8.24 THEOREM

There is a complete Boolean algebra $\mathbb{A}$ such that

$$[\![\text{MA} + \text{NDH}]\!]^{\mathbb{A}} = 1.$$

Proof

Let $\mathbb{C}$ be the complete Boolean algebra given in 8.23. By the main theorem 8.22 and the fullness lemma, there is a term u in $V^{\mathbb{C}}$ such that

$[\![$'u is a complete Boolean algebra'$]\!]^{\mathbb{C}}$

$$= [\![\text{'}[\![\text{MA} + \text{NDH}]\!]^u = 1\text{'}]\!]^{\mathbb{C}} = 1.$$

We obtain $\mathbb{A}$ by one final appeal to 8.14(i) (with $\mathbb{C}$ in place of $\mathbb{B}$). ∎

We have now reached the goal of this book.

8.25 THEOREM

Assume that $\mathfrak{M}$ is a model such that $\mathfrak{M} \models$ ZFC. Then there is a model $\mathfrak{N}$ extending $\mathfrak{M}$ such that $\mathfrak{N} \models$ ZFC + MA + NDH. ∎

8.26 THEOREM

Assume that ZFC is consistent. Let NDH be the sentence: *for each compact space* X, *each homomorphism from* C(X,$\mathbb{C}$) *into a Banach algebra is continuous.* Then NDH is independent from ZFC.

Proof

Assume that there is a model of ZFC. By 8.25,

there is a model of ZFC + NDH. By 7.38, there is a model of ZFC + CH; by 1.10, this model is a model of ZFC + ¬NDH.

By Gödel's completeness theorem, NDH is independent of ZFC. ∎

8.27 NOTES

As we remarked, the initial development of iterated forcing is due to Solovay and Tennenbaum ([67]), who first showed that MA + ¬CH is relatively consistent with ZFC. Our iteration theorems 8.12 and 8.14 are essentially given in [67, §5], where the following notation is used. Let $\mathfrak{B}$ be a complete Boolean algebra, and let $\mathfrak{D}$ be a complete Boolean algebra in $V^{\mathfrak{B}}$ (corresponding to our u). Then the complete Boolean algebra corresponding to $\mathfrak{C}$ in 8.14 is denoted by $\mathfrak{B} \tilde{\otimes} \mathfrak{D}$: its relation to $\mathfrak{B} \otimes \mathfrak{D}$, the tensor product of $\mathfrak{B}$ and $\mathfrak{D}$ is discussed. Infinite iterations are discussed in [67, §6]: our account is quite close to this, but our unions replace certain direct limits. The construction of a complete Boolean algebra $\mathfrak{B}$ such that $[\![\mathrm{MA}]\!]^{\mathfrak{B}} = 1$ is given in [67, §7]. Iterated forcing is discussed in [49, Chapter VIII], for example.

The proof of Theorem 8.22 (from [72]) is a development of an argument of Kunen, who proved that "There are no $(\mathfrak{c},\mathfrak{c})$-gaps in $(\mathbb{N}^{\mathbb{N}}, <_F)$" is relatively consistent with ZFC + MA.

Since NDH is relatively consistent with MA + ¬CH, and since CH implies ¬NDH, the construction of discontinuous homomorphisms from algebras C(X) (for each infinite, compact space X) is one that cannot be carried through in the more general theory ZFC + MA. In fact, we *conjecture* that NDH is actually independent of MA + ¬CH: it is shown in [72] that it is consistent with MA + ¬CH that $(\mathbb{R},<)$ does embed in $(\mathbb{N}^{\mathbb{N}}, <_F)$, and so the existence of such an embedding is independent of MA + ¬CH.

We have shown that NDH is relatively consistent with $\mathrm{MA} + 2^{\aleph_0} = \aleph_2$. In fact, it can be arranged that $2^{\aleph_0}$ can take any value "consistent with MA".

BIBLIOGRAPHY

[1] E. Albrecht and M. Neumann, 'Automatic continuity of generalized local linear operators', *Manuscripta Math.*, 32 (1980), 263-294.

[2] W.G. Bade and P.C. Curtis, Jr., 'Homomorphisms of commutative Banach algebras', *Amer. J. Math.*, 82 (1960), 589-608.

[3] J. Barwise, 'An introduction to first-order logic', in [4], 5-46.

[4] J. Barwise (Ed.). *Handbook of mathematical logic*, North Holland, Amsterdam, 1977.

[5] J.E. Baumgartner, 'Applications of the proper forcing axiom', in [50], 913-959.

[6] J.L. Bell, *Boolean-valued models and independence proofs in set theory*, Oxford University Press, 1977; second edition, 1985.

[7] J.L. Bell and A.B. Slomsom, *Models and ultra-products: an introduction*, North Holland, Amsterdam, 1969.

[8] F.F. Bonsall and J. Duncan, *Complete normed algebras*, Springer-Verlag, New York, 1973.

[9] J.P. Burgess, 'Forcing', in [4], 403-452.

[10] C.C. Chang and H.J. Keisler, *Model theory*, North Holland, Amsterdam, 1973.

[11] P.J. Cohen, 'The independence of the continuum hypothesis I', *Proc. Nat. Acad. Sci. USA*, 50 (1963) 1143-1148.

[12] P.J. Cohen, 'The independence of the continuum hypothesis II', *Proc. Nat. Acad. Sci. USA*, 51 (1964), 105-110.

[13] P.J. Cohen, *Set theory and the continuum hypothesis*, Benjamin, New York, 1966.

[14] W.W. Comfort and S. Negrepontis, *The theory of ultrafilters*, Springer-Verlag, Berlin, 1974.

[15] H.G. Dales, 'Automatic continuity: a survey', *Bull. London Math. Soc.*, 10 (1978), 129-183.

[16] H.G. Dales, 'A discontinuous homomorphism from C(X)', *Amer. J. Math.*, 101 (1979), 647-734.

[17] H.G. Dales and J.R. Esterle, Monograph in preparation.

[18] H.G. Dales and R.J. Loy, 'Prime ideals in algebras of continuous functions', *Proc. Amer. Math. Soc.*, 98 (1986), 426-430.

[19] M. Davis, 'Hilbert's tenth problem is unsolvable', *Amer. Math. Monthly*, 80 (1973), 233-269.

[20] M. Davis, Yu.V. Matijasevič and J. Robinson, 'Hilbert's tenth problem. Diophantine equations: positive aspects of a negative solution', *Proc. Symposium on Hilbert's problem*, American Math. Soc., 1975.

[21] K.J. Devlin and H. Johnsbråten, *The Souslin problem*, Lecture Notes in Maths. 405, Springer-Verlag, Berlin, 1974.

[22] A. Dow, 'On ultrapowers of Boolean algebras', *Topology Proc.*, 9 (1984), 269-291.

[23] H.B. Enderton, *A mathematical introduction to logic*, Academic Press, New York, 1972.

[24] J.R. Esterle, 'Solution d'un problème d'Erdös, Gillman et Henriksen et application à l'étude des homomorphismes de C(K)', *Acta Math. Acad. Sci. Hungar.*, 30 (1977), 113-127.

[25] J.R. Esterle, 'Semi-normes sur C(K)', *Proc. London Math. Soc.*, (3) 36 (1978), 27-45.

[26] J.R. Esterle, 'Sur l'existence d'un homomorphisme discontinu de C(K)', *Proc. London Math. Soc.*, (3) 36 (1978), 46-58.

[27] J.R. Esterle, 'Injection de semi-groupes divisibles dans des algèbres de convolution et construction d'homomorphismes discontinus de C(K)', *Proc. London Math. Soc.*, (3) 36 (1978), 59-85.

[28] J.R. Esterle, 'Homomorphismes discontinus des algèbres de Banach commutatives separables', *Studia Math.*, 66 (1979), 119-141.

[29] J.R. Esterle, 'Theorems of Gelfand-Mazur type and continuity of epimorphisms from C(K)', *J. Funct. Anal.*, 36 (1980), 273-286.

[30] J.R. Esterle, 'Universal properties of some commutative radical Banach algebras', *J. Reine Angew. Math.*, 321 (1981), 1-24.

[31] J.R. Esterle, 'Elements for the classification of commutative radical Banach algebras', *Radical Banach algebras and automatic continuity*, Lecture Notes in Maths. 975 (1983), 4-65.

[32] M. Foreman, M. Magidor and S. Shelah, 'Martin's Maximum, saturated ideals, and non-regular ultrafilters, Part I', *Ann. of Math.*, to appear.

[33] M. Foreman and W.H. Woodin, 'GCH can fail everywhere', in preparation.

[34] D.H. Fremlin, *Consequences of Martin's Axiom*, Cambridge Tracts in Mathematics 84, Cambridge University Press, 1984.

[35] T.W. Gamelin, *Uniform algebras*, Prentice-Hall, Englewood Cliffs, New Jersey, 1969.

[36] L. Gillman and M. Jerison, *Rings of continuous functions*, van Nostrand Reinhold, New York, 1960.

[37] K. Gödel, *The consistency of the axiom of choice and of the generalized continuum hypothesis with the axioms of set theory*, Ann. of Math. Studies 3, Princeton University Press, Princeton, 1940.

[38] P.R. Halmos, *Naïve set theory*, van Nostrand Reinhold, New York, 1960.

[39] P.R. Halmos, *Lectures on Boolean algebras*, van Nostrand, New York, 1963.

[40] F. Hausdorff, 'Summen von $\aleph_1$ Mengen', *Fund. Math.*, 26 (1936), 241-255.

[41] N. Jacobson, *Basic algebra II*, W.H. Freeman, San Francisco, 1980.

[42] T. Jech, 'Non-provability of Souslin's hypothesis', *Comment. Math. Univ. Carolin.*, 8 (1967), 291-305.

[43] T. Jech, *The axiom of choice*, North Holland, Amsterdam, 1973.

[44] T. Jech, *Set theory*, Academic Press, New York, 1978.

[45] T. Jech, *Multiple forcing*, Cambridge Tracts in Mathematics 88, Cambridge University Press, 1987.

[46] B.E. Johnson, 'Continuity of homomorphisms of algebras of operators', *J. London Math. Soc.*, 42 (1967), 537-541.

[47] B.E. Johnson, 'Norming C(Ω) and related algebras', *Trans. Amer. Math. Soc.*, 220 (1976), 37-58.

[48] I. Kaplansky, 'Normed algebras', *Dake Math. J.*, 16 (1949), 399-418.

[49] K. Kunen, *Set theory, an introduction to independence proofs*, North Holland, Amsterdam, 1980.

[50] K. Kunen and J.E. Vaughan (Ed.), *Handbook of set-theoretic topology*, North Holland, Amsterdam, 1984.

[51] K.B. Laursen, 'Automatic continuity of generalized intertwining operators', *Dissertationes Math. (Rozprawy Mat.)*, 189, 1981.

[52] D.A. Martin and R.M. Solovay, 'Internal Cohen extensions', *Ann. Math. Logic*, 2 (1970), 143-178.

[53] G.H. Moore, *Zermelo's axiom of choice: its origins, development, and influence*, Springer-Verlag, New York, 1982.

[54] S. Priess-Crampe, *Angeordnete Strukturen: Gruppen, Körper, projektive Ebenen*, Springer-Verlag, Berlin, 1983.

[55] H. Rasiowa and R. Sikorski, *The mathematics of metamathematics*, Monografie Matematyczne 41, (Warsaw, 1963).

[56] J.G. Rosenstein, *Linear orderings*, Academic Press, New York, 1982.

[57] M.E. Rudin, Martin's Axiom, in [4], 491-501.

[58] W. Rudin, *Functional Analysis*, McGraw-Hill, New York, 1973.

[59] D. Scott, 'A proof of the independence of the Continuum Hypothesis', *Math. Systems Theory*, 1 (1967), 89-111.

[60] D. Scott, *Lectures on Boolean-valued models of set theory*, Lecture Notes of the UCLA Summer Institute on Set Theory, 1967.

[61] S. Shelah, *Proper forcing*, Lecture Notes in Mathematics 940, Springer-Verlag, Berlin, 1982.

[62] J.R. Shoenfield, *Mathematical Logic*, Addison-Wesley, Reading, Mass., 1967.

[63] J.R. Shoenfield, 'Martin's Axiom', *Amer. Math. Monthly*, 82 (1975), 610-617.

[64] A.M. Sinclair, Homomorphisms from $C_o(\mathbb{R})$', *J. London Math. Soc.*, (2) 11 (1975), 165-174.

[65] A.M. Sinclair, *Automatic continuity of linear operators*, London Math. Soc. Lecture Note Series 21, Cambridge University Press, 1976.

[66] C. Smorynski, 'The incompleteness theorems', in [4], 821-866.

[67] R.M. Solovay and S. Tennenbaum, 'Iterated Cohen extensions and Souslin's problem', *Ann. of Math.*, 94 (1971), 201-245.

[68] K.D. Stroyan and W.A.J. Luxemburgh, *Introduction to the theory of infinitesimals*, Academic Press, New York, 1976.

[69] R.C. Walker, *The Stone-Čech compactification*, Springer-Verlag, Berlin, 1974.

[70] W. Weiss, 'Versions of Martin's Axiom', in [50], 827-886.

[71] E. Wimmers, 'The Shelah P-point independence theorem', *Israel J. Math.*, 43 (1982), 28-48.

[72] W.H. Woodin, 'Set theory and discontinuous homomorphisms from Banach algebras', *Mem. Amer. Math. Soc.*, to appear.

INDEX OF NOTATION

INDEX